MICROCOSMOS
THE WORLD OF ELEMENTARY PARTICLES

Fictional Discussions between
Einstein, Newton, and Gell-Mann

MICROCOSMOS

THE WORLD OF ELEMENTARY PARTICLES

Fictional Discussions between
Einstein, Newton, and Gell-Mann

Harald Fritzsch

Ludwig Maximilian University of Munich, Germany

Translated by

Harald Fritzsch and Jeanne Rostant

World Scientific

NEW JERSEY · LONDON · SINGAPORE · BEIJING · SHANGHAI · HONG KONG · TAIPEI · CHENNAI

Published by

World Scientific Publishing Co. Pte. Ltd.

5 Toh Tuck Link, Singapore 596224

USA office: 27 Warren Street, Suite 401-402, Hackensack, NJ 07601

UK office: 57 Shelton Street, Covent Garden, London WC2H 9HE

Library of Congress Cataloging-in-Publication Data
Fritzsch, Harald, 1943–
  [Mikrokosmos. English]
  Microcosmos--the world of elementary particles : fictional discussions between Einstein,
Newton, and Gell-Mann / Harald Fritzsch, Ludwig Maximilian University Munich, Germany.
    pages cm
    ISBN 978-9814449984 (hardcover : alk. paper)
    1. Particles (Nuclear physics)--Popular works. I. Title.
    QC793.26.F75413 2013
    539.7'2--dc23
                                    2013015752

British Library Cataloguing-in-Publication Data
A catalogue record for this book is available from the British Library.

Mikrokosmos © 2012 by Piper Verlag GmbH, München

In-house Editor: Alan Tew Tiong Phew

Typeset by Stallion Press
Email: enquiries@stallionpress.com

Printed in Singapore

CONTENTS

PREFACE

The development of the sciences, especially of physics, is strongly connected with new insights into the microscopic structure of matter. At the beginning of the 19th century, physicists and chemists came to the conclusion that matter consists of atoms (the smallest units of matter). The first experimental indications towards an atomic structure were obtained at the beginning of the 20th century.

Today, we know that atoms consist of atomic nuclei and electrons (the electrons are found in the region around the nuclei). Nuclei, in turn, are bound states of protons and neutrons, which are collectively called nucleons. It was assumed that protons and neutrons are elementary particles, like electrons, but it turns out that nucleons have an internal structure: a nucleon consists of three quarks. In contrast, electrons and quarks seem to be truly elementary particles. As of today, no internal structure of electrons has been found. According to experiments, electrons and quarks are point-like. If, however, they do have an internal structure, their diameters must be less than 10^{-17} cm.

The physics of elementary particles started around 1950 as a special field of nuclear physics. 25 years later, the Standard Model of particle physics was developed. In this model, there are particles of matter, such as electrons and quarks, and particles that mediate forces, such as photons (which mediate the electromagnetic force).

The electron is a particle that has five relatives, which are called leptons. In 1936, the muon (a charged lepton) was discovered. Its mass is about 200 times larger than the mass of the electron. In 1975, another charged lepton, the tauon, was found. Its mass is about 3500 times larger than that of the electron. Besides the three charged leptons (namely, the electron, the muon, and the tauon), there are three neutral leptons (and they are called neutrinos).

In our Universe, there are six different quarks. Among them are the "up" quark (u) and the "down" quark (d), which are the constituents of the nucleons. The other quarks are the "strange" quark (s), the "charm" quark (c), the "bottom" quark (b), and the "top" quark (t). Stable matter in the Universe consists of "up" and "down" quarks (the building blocks of protons and neutrons). The other quarks are constituents of unstable particles.

The strong force between quarks are mediated by massless gluons, and the weak force are mediated by very massive weak bosons. The electromagnetic and weak forces are described by the theory of the electroweak interaction. The strong interaction between quarks and gluons are described by the theory of quantum chromodynamics (QCD).

The Standard Model is the combination of the theory of the electroweak interaction and the theory of quantum chromodynamics. This model is able to describe almost all the observed particle-physics phenomena.

In the Standard Model, there are 28 constants, which have to be determined by experiments. Among them are the 12 masses of leptons and quarks. Nobody knows the origin of these constants. Many physicists believe that the Standard Model is not the final theory of our Universe, but only a good approximation, because with a final theory, one should be able to calculate at least some of these fundamental constants.

Particle physicists explore the internal structure of matter in our Universe. Unfortunately, this kind of research is not cheap (modern accelerators, such as the Large Hadron Collider (LHC) at CERN,[a] cost several billion Euros) and the question of whether funding such research is worthwhile arises. While there are no direct applications of particle physics to the industry, indirect applications, such as the invention of the internet and the World Wide Web at CERN in 1990, have turned out to be rather important.

This book describes fictional discussions on the topic of particle physics between Albert Einstein, Isaac Newton, Murray Gell-Mann, and the modern physicist Adrian Haller. By following these discussions, the reader will get an overview of the present status of particle physics and come to understand why particle physics is an exciting field.

Harald Fritzsch
Munich, July 2013

---

[a] CERN is the European Organization for Nuclear Research. It stands for *Conseil Européen pour la Recherche Nucléaire* (French).

A Meeting in Pasadena

Adrian Haller, a physics professor at the University of Bern, Switzerland, was flying with the airline Swiss from Zurich to Los Angeles. Shortly after take-off, he could see the mountains of the Bernese Oberland, especially the Finsteraarhorn, the Eiger and the Jungfrau and, further away, the Matterhorn near Zermatt.

The airplane flew towards Paris, and soon afterwards, it was above the Atlantic Ocean. Haller started to do some work, but after some time, he got tired and took a nap...

After crossing the mountains of San Gabriel, north of Pasadena, the airplane landed at the international airport of Los Angeles. After collecting his luggage, Haller took a taxi to the Athenaeum, the guesthouse of the California Institute of Technology (Caltech), Pasadena.

At the reception desk, Haller asked for the room that Einstein had used in 1929, since Haller stayed in that room every time he visited Caltech. But the receptionist told him that this time he would have to change to another room since the "Einstein room" was taken by someone else.

"Who is in this room now?" asked Haller.

"You will be astonished, but it is a Mr. Einstein who arrived yesterday. But I don't know whether this person has anything to do with the famous Einstein. He looks a bit like him, so it could be his son. He asked to stay in the room where Einstein once stayed, and since his name was Einstein, I agreed. I assumed that you would not mind being in another room this time.

You will see Mr. Einstein soon, since he has already asked me several times whether you have arrived. In the Athenaeum, there are two other gentlemen whom you know: an Englishman called Isaac Newton, and Professor Murray Gell-Mann, whom I know well, since he was a professor at Caltech until 1993."

"Good, then, I will see them soon," responded Haller.

Haller went up to his room. After unpacking his luggage, he went to the Einstein room and knocked on the door. Albert Einstein opened the door.

Einstein: Hello, Mr. Haller. You have finally arrived here in Southern California. Welcome.

Haller: Greetings, Mr. Einstein. I've just realized that you are staying in the room that you once stayed in 1929. The lady at the reception thought that you might be the son of the famous Einstein. She did not realize that you are the real Albert Einstein!

Einstein: Yes, she must have assumed that I have been dead since 1955. It is nice for me to come back here to Southern California, which changes every day, but the Athenaeum remains as it was in 1929. Even the restaurant downstairs has not changed.

Isaac Newton and Murray Gell-Mann have also arrived. I've already had a long conversation with Gell-Mann. I have known him since we were in Princeton. I remember that we used to speak a mixture of English and German whenever we met at the Institute for Advanced Study. When I met him, I would say, "Guten Morning, Mr. Gell-Mann," and he would respond, "Good Morgen, Herr Einstein." And of course, I know that Gell-Mann has become the leading physicist in particle physics. I suggest that we meet him and Newton in about one hour's time downstairs in the restaurant.

Haller went back to his room. He had to think about the discussion on particle physics that was starting soon. After one hour, he went down to the restaurant, where Einstein, Gell-Mann, and Newton were already sitting at a table in a corner. Haller shook hands with Isaac Newton and Murray Gell-Mann. He had known Newton from previous discussions, and he had collaborated with Gell-Mann for many years, notably from 1970 until 1976.

Haller: Dear Gentlemen, I am glad that we are meeting now here in the Athenaeum. Before we start our discussions on elementary particles, however, we should have our dinner. Let me order the excellent Filet Mignon for all of us.

The others agreed, and soon the Filet Mignon was served, together with a Cabernet Sauvignon from the Napa Valley. After an excellent dinner, they had coffee in the Athenaeum's small library behind the restaurant.

Einstein knew this room well, since he had given a lecture on his general theory of relativity in this library in 1929, soon after he was informed about the observation of Edwin Hubble concerning the flight of distant galaxies (which turns out to be the first signal of the Big Bang).

AN INTRODUCTION TO PARTICLE PHYSICS

The next morning, the discussion started in the library of the Athenaeum.

Haller: Gentlemen, the topic of our discussion during the coming days will be on the topic of particle physics. I think it would be a good idea if Murray Gell-Mann could start off with a general introduction.

Gell-Mann: Well, particle physics is an offspring of atomic and nuclear physics. Let me first make a few remarks about atomic physics. The first elementary particle to be discovered is the electron. It was discovered in 1897 by J. J. Thomson at the Cavendish laboratory. Furthermore, it is the lightest charged particle, with an electric charge of $-e$ (where e is the elementary electric charge) and a mass of about 0.511 MeV. According to Einstein's mass–energy relation, this mass corresponds to 9.109×10^{-31} kg.

Electrons are constituents of atoms. Thus, atoms have a substructure. Around 1900, it was assumed that electrons move around inside atoms. In 1911, Ernest Rutherford, who was then working at the University of Manchester, made an interesting discovery. Rutherford investigated the internal structure of atoms using alpha particles, which are emitted during certain radioactive decays. He found that an alpha particle after going through a thin metal foil sometimes underwent a sudden change in its direction of flight. Rutherford interpreted this behavior as a consequence of a collision between the alpha particle and a compact, positively charged object inside the metal atom; he called the positively charged object "atomic nucleus".

Thus, inside an atom, there is a very small atomic nucleus, which is positively charged and surrounded by a cloud of electrons. The electric charge of a nucleus is equal in strength but opposite in sign to the total electric charge of its surrounding electrons. Thus, atoms are neutral. In addition, the size of an atom is determined by the size of the cloud of its surrounding electrons.

Ernest Rutherford, who discovered that atoms are not elementary, but consist of electrons and nuclei

Rutherford measured the radius of an atomic nucleus using alpha rays and found that it was about 10 000 times smaller than the radius of the atom. This means that if we blew up an atom to the size of a balloon with a radius of 10 m, its atomic nucleus would have a radius of only 1 mm. Moreover, almost the entire mass of an atom is given by the mass of its nucleus.

Rutherford found out that the nucleus of the lightest atom, the hydrogen atom, consists of only one particle — the proton — and that the nucleus is surrounded by only one electron. The proton has a positive charge that is equal in strength but opposite in sign to the electron charge; the proton has a mass that is about 2000 times larger than the electron mass (mass$_{proton}$ = 9383 MeV).

The helium atom has two electrons, and in its nucleus, there are two protons. The iron atom has 26 electrons; the uranium atom, 92 electrons.

Newton: Question: Why, then, are all hydrogen atoms identical? If a hydrogen atom consists of an electron moving around a proton, there should be many different hydrogen atoms: small ones and larger ones, depending on the radius of the trajectory of the electron. Why, then, are all hydrogen atoms identical? I cannot understand this.

Gell-Mann: Indeed, if atoms were classical systems, there would be an infinite number of different atoms, and it would not be possible to specify the mass of an atom. However, atoms are not classical systems, but systems described by quantum physics. Hydrogen atoms are normally in the state with the lowest energy, the ground state. Thus, they are all identical: a hydrogen atom here on earth is identical to a hydrogen atom in a distant star.

Initially, it was assumed that an atomic nucleus is a bound state of protons. No one understood how such a nucleus could be stable since it was known from observation that protons repel one another. Also, the mass of a nucleus was a mystery. For example, the helium nucleus thought of as a bound state of two protons should have a mass that is twice the mass of a proton, but it turns out that its mass is about four times the mass of a proton.

Rutherford worried about this problem, and in 1919, he came to the conclusion that (1) an atomic nucleus does not consist of protons alone, but there is another neutral particle, which he called a neutron, inside a nucleus, and (2) the neutron mass should be similar to the proton mass.

Rutherford is right. In 1932, the neutron was discovered by James Chadwick. Its mass is about 9396 MeV, slightly larger than the mass of a proton.

With this discovery, the mass of atomic nuclei can now be understood. The helium nucleus not only has two protons, but also two neutrons, and for this reason, the mass of a helium nucleus is four times the mass of a hydrogen nucleus. On a side note, alpha particles, which were used by Rutherford in his experiments, are actually helium nuclei.

Newton: I read recently that electrons and protons have a new quantum number — spin — which is something like an angular momentum. I find this rather peculiar since an electron is a point-like particle, without any extension. Where, then, could this angular momentum come from? Something should be moving around inside an electron, but if so, an electron cannot be point-like.

Haller: Indeed, spin is a peculiar quantum number. It was discovered in 1922 by Otto Stern and Walter Gerlach when they were investigating silver atoms using a strong magnetic field. They observed that a ray of silver atoms after passing through a strong magnetic field splits up into two rays.

Since a silver atom has only one electron in its outer shell, it implies that, essentially, the Stern–Gerlach experiment investigated the properties of a single electron. (We know today that a ray of electrons after passing through a strong magnetic field will split up into two. But in 1922, physicists could not produce a ray of electrons.)

Fig. 2.1. *An electron with its spin, which is either positive or negative*

Newton: This is strange. Since electrons are charged, point-like particles, they are all identical. Consequently, I would expect a ray of electrons not to be split up by a magnetic field.

Gell-Mann: No, remember spin, which you mentioned a while ago. Electrons have this new quantum number, which was introduced and named as such by Wolfgang Pauli around 1925. It was initially thought to be something like an angular momentum and is described by two numbers: $\pm\frac{1}{2}\hbar$. (Here, $\hbar$ is the reduced Planck constant,[a] the fundamental constant of quantum physics.) Spin is a quantum phenomenon, and its significance disappears in the classical limit. Thus, spin is not an angular momentum, which is related to the rotation of a particle. The point-like electron does not have an angular momentum, but it has a spin.

Let us consider an electron at rest. Its spin in the z-direction will either be $+\frac{1}{2}$ or $-\frac{1}{2}$ (the constant $\hbar$ is set as one). These two states are described in Fig. 2.1. When electrons move through a magnetic field,

Wolfgang Pauli, one of the pioneers of quantum mechanics

[a]The reduced Planck constant is the Planck constant divided by 2π.

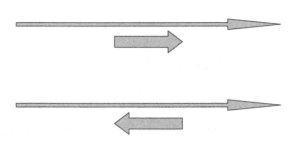

Fig. 2.2. *A moving electron is described by an arrow for its momentum and a second (smaller) arrow for its spin. If the momentum arrow and the spin arrow are pointing in the same direction, the electron is said to be right-handed; if the two arrows are pointing in the opposite directions, the electron is said to be left-handed. In general, an electron is a superposition of these two possible states.*

50% of the electrons will have their spins parallel to the field, and the remaining 50% will have their spins anti-parallel to it (see Fig. 2.2). Thus, a ray of electrons will be split up into two by the magnetic force, as observed by Stern and Gerlach.

Haller: Every particle has a certain spin, which may be zero. For example, pions, which we shall discuss later, have spin 0. Protons and neutrons, like electrons, have spin $\frac{1}{2}$. Photons (the particles of light) have spin 1. There also exist particles with spin $\frac{3}{2}$, but they are unstable, as are particles with spin 2.

Einstein: In quantum mechanics, physical systems are described by a differential equation that was introduced by Erwin Schrödinger. The Schrödinger equation, however, is an equation in which spin does not appear. Which equation, then, can be used to describe spin?

Haller: In the early 1920s, Paul Dirac, an electrical engineer who was interested in physics, dwelled on this problem in Cambridge, England. In 1928, he found the solution — an equation that is now called the Dirac equation.

Dirac wanted to find a relativistic version of the Schrödinger equation. The Schrödinger equation is non-relativistic since it has second derivatives with respect to the space coordinates but only a first (and no

Paul Dirac

second) derivative with respect to time: space and time are not unified as spacetime in the equation as demanded by Einstein's theory of relativity.

One possible way to obtain a relativistic equation is to substitute the first derivative with respect to time by a second derivative with respect to time. This had been tried, in particular by Oscar Klein in Stockholm. The resultant equation is called the Klein–Gordon equation. However, it has several problems. Among them, I just want to mention that probability is not conserved. Dirac tried another possibility: he replaced the second derivatives with respect to space with the first derivatives (with respect to space).

Newton: I do not understand. The first derivative with respect to space is a gradient. It is a vector, which points in a specific direction. This is not the case for the second derivatives with respect to space. The sum of the three second derivatives with respect to space is a scalar, and it is for this reason that it was used by Schrödinger.

Gell-Mann: You are right, but Dirac did not consider a normal gradient. He multiplied his derivatives with 4×4 matrices — three dimensions for the three dimensions of space and one dimension for time. These 4×4 matrices are now called Dirac matrices. Since Dirac used 4×4 matrices, the wave function of a spin-$\frac{1}{2}$ particle must be described by four wave functions.

Newton: But why four wave functions? I thought an electron, because of its spin, has only two wave functions.

Gell-Mann: Two components, indeed, describe spin. The other two components describe another particle. According to Dirac, a particle like the electron has an associated particle, an antiparticle as he called it. A particle and its antiparticle have the same mass but opposite electric charges.

When Dirac introduced his equation in 1928, he had no idea why antiparticles had not been observed in experiments. He tried to consider protons as the antiparticle of electrons, but this does not work since a proton has a different mass from an electron. In 1931, Dirac came to the conclusion that the antiparticle of electrons must exist.

In 1932, Carl David Anderson, a physics professor at Caltech, Pasadena, during a detailed investigation of cosmic rays, found in his cloud chamber a particle that makes a trace like the one made by an electron, but the curvature of its trace is opposite that of an electron (see Fig. 2.3). Thus, it must have roughly the same mass as an electron, but have the opposite electric charge. This positively charged particle was later called a "positron".

Haller: Today, we know that the existence of the positron is not directly related to the Dirac equation. In fact, every particle has its own antiparticle. Interestingly, some neutral particles, such as photons, are their own antiparticles. A particle and its antiparticle are related by a symmetry called C-symmetry; C stands for charge conjugation, and as a transformation, it refers to the transition from a particle to its antiparticle. One might think that this symmetry is an exact symmetry of nature, but it turns out that C-symmetry is violated by the weak interaction. This, we shall discuss later.

Carl David Anderson, who discovered the positron

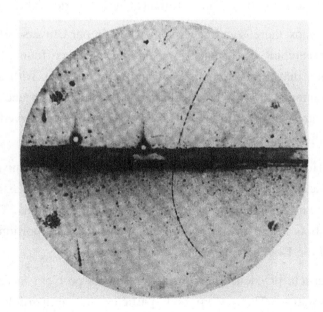

Fig. 2.3. *The trace of a positron in Anderson's cloud chamber*

When a positron and an electron collide, they annihilate each other and produce, in most cases, two — sometimes, three — photons in a process that conserves energy. If both particles are nearly at rest, their masses turn into the energy of the photons according to Einstein's mass–energy relation. The reverse reaction can also happen: if two photons collide and their energies are large enough, an electron and a positron are generated. In other words, matter and antimatter can be generated from light!

In the early 1950s, the antiparticle of the proton, the antiproton, was discovered using an accelerator in Berkeley, California.

Einstein: Since there is an antiparticle for every particle, the following question naturally arises: why is our world not symmetrical between particles and antiparticles? With antiparticles, one can build atoms — actually, anti-atoms — and they form a new kind of matter — the antimatter. But I do not see any antimatter in our world. Where is it?

Haller: Perhaps, there is antimatter somewhere in our Universe. Physicists have been searching for it, but so far, nothing has been found. Suppose, for example, that a star composed of antimatter collides with a normal star. The matter of the two stars would annihilate each other and generate a lot of electromagnetic radiation. Because we have not observed such an event, we think that there is no antimatter in our Universe.

Einstein: Hold on, I have a problem. It is assumed that our Universe was created in a Big Bang, in which a huge amount of energy was released, and matter and antimatter were created in equal proportions. Thus, our Universe should have matter and antimatter in equal proportions unless the idea of the Big Bang is wrong.

Haller: Particle physicists have investigated this problem in detail and found a solution. There were both particles and antiparticles in the Universe immediately after the Big Bang. But C-symmetry was slightly broken, and this resulted in a very small surplus of particles: there were ten billion and one particles for every ten billion antiparticles. The ten billion antiparticles annihilated with ten billion particles, and one particle remains. Over time, these leftover particles build the matter in our Universe, and there is no antimatter.

I should also mention that this explanation is correct only if a proton is not stable, but decays, for example, into a positron and a photon. Later, we shall see that the decay of a proton is also necessary in some theories that describe a grand unification of all the forces in nature.

Einstein: If protons do decay, we should observe them experimentally. Are there any such experiments?

Haller: Yes, there are. In fact, physicists have built huge particle detectors and installed them in mines (so as to avoid cosmic rays). The world's largest detector, Kamiokande, is located south of Toyama, Japan, near the small village Kamioka (see Fig. 2.4). Thus far, no proton decay has been detected. If protons do decay, they decay very slowly: the present lower limit on the lifetime of a proton is about 10^{32} years.

Let me also mention that a particle can either have an integer spin, that is, spin 0, 1, 2, and so on, or a half-integer spin, that is, spin $\frac{1}{2}$, $\frac{3}{2}$, and so on. A particle with an integer spin is called a boson; a particle with a half-integer spin is called a fermion.

The lightest meson,[b] a type of particle, is the pion (also known as the π-meson). It has spin 0 and is, therefore, a boson. It is unstable: it decays into an electron and a neutrino, a type of neutral fermions. Neutrinos are peculiar particles: they do not interact via the electromagnetic force — since they are neutral — but only via the weak interaction (which we shall discuss in more detail later). For this reason, neutrinos

---

[b] A meson is any particle that is a bound state of one quark and one antiquark.

Fig. 2.4. *Kamiokande, the world's largest particle detector, in Japan*

can propagate through matter over long distances: a neutrino can easily fly through the Earth.

Neutrinos were discovered in 1955 at the Savannah River Site, a nuclear reservation in South Carolina. As you might have guessed, a nuclear reactor emits lots of neutrinos. When a neutrino collides with an atomic nucleus, it turns into an electron, which had already been discovered in 1955 (it is because of this reaction of neutrinos that they were discovered).

Gell-Mann: Let us return to the topic of atoms. The simplest atom is the hydrogen atom: it has one proton as its nucleus and one electron in its

MICROCOSMOS: THE WORLD OF ELEMENTARY PARTICLES

shell. The next simplest atom is the helium atom: it has two protons and two neutrons as its nucleus and two electrons in its shell. Then, we have the lithium atom: it has three protons and four neutrons as its nucleus and three electrons in its shell. Lastly, let us mention the uranium atom: it has 92 protons and 146 neutrons as its nucleus and 92 electrons in its shell.

In 1932, Werner Heisenberg wondered why the mass of the newly discovered neutron is about the same as the mass of a proton. Also, the strong interaction between a neutron and an atomic nucleus is about the same strength as that between a proton and the same (for comparison) atomic nucleus. With these observations, Heisenberg introduced a new symmetry, which he called "isospin", and in doing so, he assumed that a physical system changes only very little if a proton is exchanged for a neutron.

For the first time, an internal symmetry was introduced in physics. Although theorized to be valid for the strong interaction inside atomic nuclei, the isospin symmetry is broken by the electromagnetic interaction.

By 1932, all the building blocks of normal matter (namely, the electron, the proton, and the neutron) were discovered. Then, in 1936, a new charged particle called the muon was discovered in cosmic rays. It has a mass of about 106 MeV, that is, nearly 200 times larger than the mass of an electron. To date, we do not know why muons exist; they are not needed in the formation of normal matter.

Haller: Muons are produced in nuclear collisions in the upper atmosphere, about 100 km above the surface of the Earth. Then, when they arrive at the surface, they can be investigated. Muons are not stable; the lifetime of a muon is about 2×10^{-6} s. A muon decays into an electron and two neutrinos shortly after its production in a particle collision. A type of neutrino called the electron neutrino is the neutral partner of the electron; another type called the muon neutrino is the neutral partner of the muon.

Newton: Question: You mentioned the short lifetime of a muon. Since muons are produced about 100 km above the surface of the Earth, they will decay before they reach the surface. Thus, it would not be possible to observe muons in cosmic-ray experiments at sea level.

Haller: But don't forget that these muons move at almost the speed of light. In this case, Albert's theory of relativity, which predicts the phenomenon of time dilation, comes into play. After taking this into account, there is no problem for these muons to arrive at the surface. In fact, the detection of these muons at sea level has given the first proof of time dilation.

I shall add that these muons are not produced directly; rather, they are the decay products of the pions that are produced during the collisions between cosmic rays and atomic nuclei. A type of meson, pions have spin 0 and a mass of about 140 MeV, and they interact strongly with nucleons. There are three types of pions: two charged ones and one neutral one.

A charged pion decays into a muon and a neutrino. Like the decay of a muon, the decay of a charged pion involves the weak interaction. The lifetime of a charged pion is about 3×10^{-8} s, slightly less than the lifetime of a muon. In contrast, a neutral pion has a much shorter lifetime (only about 10^{-16} s) since it decays via the electromagnetic interaction (into two photons).

Gell-Mann: Around the 1950s, a lot of new particles were discovered in cosmic rays and in laboratories with small accelerators. These particles are not stable; they decay. In particular, a new neutral baryon[c] known as the Lambda baryon (with a mass of about 1116 MeV) was found. Later, three new particles were discovered; they are the three types of Sigma baryons: two charged ones and one neutral one. Their masses (about 1193 MeV) are larger than the mass of a Lambda particle. Then, another two new particles were discovered; they are the two types of Xi baryons: one negatively charged one and one neutral one. Their masses are about 1318 MeV. These six particles are called "hyperons".[d]

---

[c]A baryon is any particle that is a bound state of three quarks.

[d]A hyperon is any baryon that consists of at least one strange quark, but no charm or bottom quarks.

Besides the hyperons, four types of K mesons were discovered. One of them is positively charged, and another is neutral; these two mesons and their antiparticles form the four types of K mesons. Their masses are all just below 500 MeV.

Newton: Why are there three pions but four K mesons?

Haller: That is an interesting question. It has to do with the internal structure of K mesons. In 1964, Gell-Mann found the answer with his quark model, which we shall discuss later.

Newton: A Lambda hyperon is also unstable. How does it decay?

Haller: The dominant mode of decay is the one that produces a nucleon and a pion. Again, it is due to the weak interaction.

Gell-Mann: The production of hyperons and K mesons in a particle collision is rather strange: a hyperon is always produced with a K meson, never with a pion. Nobody understood why.

In the early 1950s, I was working as a postdoc in Enrico Fermi's group in Chicago. In order to explain the production of hyperons in association with K mesons, I introduced a new quantum number, which I call "strangeness", denoted by S. Nucleons and pions have $S = 0$; Lambda hyperons and Sigma hyperons have $S = -1$; Xi hyperons have $S = -2$; positively charged K mesons and neutral K mesons have $S = +1$. By postulating that the quantum number strangeness is conserved by the strong interaction, I was able to explain why hyperons are always produced together with K mesons.

Einstein: But strangeness is not conserved by the weak interaction, since, for example, a Lambda hyperon decays into a pion and a nucleon.

Gell-Mann: Yes, in the case of the weak interaction, strangeness is, indeed, not conserved. Later, we shall see that the conservation of strangeness by the strong interaction follows directly from the quark model.

Haller: Now that it is lunch time, I propose we take our lunch in the Athenaeum's restaurant.

FROM ISOSPIN TO SU(3)

3

After having had lunch in the restaurant of the Athenaeum, Einstein suggested that they drive to the top of Mount Wilson to visit its observatory (see Fig. 3.1). He had been there several times in the 1920s together with Edwin Hubble, who was a staff member there.

They drove onto the Angeles Crest Highway, and after about an hour's drive through the San Gabriel Mountains, they were at the top of Mount Wilson. There, Gell-Mann showed his colleagues its old observatory, which is now a museum.

Einstein mentioned that he was in the observatory together with Edwin Hubble in 1929. Using a 100-inch Hooker telescope (see Fig. 3.2), Hubble discovered that light coming from distant galaxies is redshifted; this redshift turns out to be an evidence for the expansion of our Universe. In addition, Hubble found that the velocity at which a distant galaxy is receding from us is proportional to its distance from us: this relation is now called "Hubble's law". When Hubble informed Einstein about this law, Einstein immediately thought that it might be an evidence for the cosmic expansion after the "Big Bang".

Fig. 3.1. *The telescope on Mount Wilson near Pasadena*

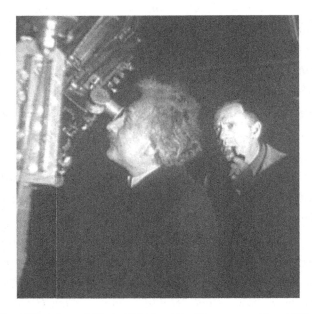

Albert Einstein and Edwin Hubble at the telescope on Mount Wilson

MICROCOSMOS: THE WORLD OF ELEMENTARY PARTICLES

Fig. 3.2. *The telescope used by Hubble for the measurement of the velocities of distant galaxies*

After visiting the observatory, they went to a nearby lookout point, and from there, they could see the city of Los Angeles, the beach of Santa Monica, and the distant island of Catalina. Then, they sat down on a bench and resumed their discussion.

Haller: Murray, in 1961, both you and Yuval Ne'eman introduced the symmetric group SU(3). Ne'eman, who was then working at the Israeli embassy in London, was trained as an engineer, but he was interested in physics and was following the lectures of Abdus Salam at Imperial College in London. Salam told Ne'eman that he should try to find a symmetry scheme to classify the new particles. Not long afterwards, Ne'eman found the symmetric group SU(3) at around the same time as you did.

Gell-Mann: Yes, Ne'eman and I were searching for a symmetry to describe the new particles, in particular the K mesons and the hyperons. I knew quite a lot about group theory, and I tried various possibilities. To me, it was clear that the symmetry we were searching for must be an extension of the isospin symmetry. However, there was a problem. The isospin symmetry is a broken symmetry, although the symmetry breaking is very small, of order 1%. It was usually assumed that the isospin symmetry is broken only by the electromagnetic interaction. In contrast, the symmetry we were searching for must be strongly broken, of order at least about 20%. Yet, there is no interaction that could break this new symmetry.

I observed that the two nucleons and the six hyperons could be described by an octet of the SU(3) group. The octet is a representation of the SU(3) group. A representation of a group, Mr. Newton, is a realization of the group by concrete elements. For example, a vector is a representation of the rotation group SO(3).

Haller: Isaac does not know anything about groups. Let me explain. Let us consider some abstract elements, say A, B, C, etc., and define a "multiplication" for these elements. These elements form a group if they satisfy some rules. First, the product of any two elements in the group is also an element in the group: $A \cdot B = C$. In the case of integer numbers, for example, we have: $2 \cdot 3 = 6$. Second, one of the elements in the group is the unit element E: $A \cdot E = E \cdot A = A$. In the case of integer numbers, the unit element is the number 1 if the "multiplication" is the normal multiplication. Note that the abstract "multiplication" may also be the normal addition: $2 + 3 = 5$. In this case, the unit element is the number 0: $A + 0 = 0 + A = A$.

Third, for every element A in the group, there also exists an inverse element A^{-1} in the group such that the product of an element and its inverse element is the unit element E: $A \cdot A^{-1} = E$. The product of any two elements in a group may be commutative: the two elements before and after the "multiplication" can be exchanged without changing the product; this is the case for integer numbers: $2 \cdot 3 = 3 \cdot 2 = 6$. This need not be the case for every group. Rather, the product is often non-commutative: $A \cdot B \neq B \cdot A$. For example, this is the case if the group elements are matrices. Gell-Mann's group SU(3) is the group of all unitary 3×3 matrices; Heisenberg's isospin group SU(2) is the group of all unitary 2×2 matrices. In both of these cases, the "multiplication" is non-commutative.

The octet of the SU(3) group contains the two nucleons, the Lambda particle, the three Sigma particles, and the two Xi particles (see Fig. 3.3). The problem is the strong breaking of SU(3) symmetry: the particles in the octet have quite different masses. If the SU(3) group were to describe an exact symmetry, the mass of the proton and the masses of the hyperons would be the same. In reality, however, the mass of the proton is about 938 MeV and the masses of the Xi particles are about 1318 MeV. Thus, the symmetry breaking is rather large, of order 20%.

Newton: I do not understand why the SU(3) group is relevant. Heisenberg introduced the SU(2) group since he wanted to describe the proton and the neutron, the two particles that form the doublet of SU(2). The nucleons and the hyperons are eight particles altogether; thus, one should consider the SU(8) group, I guess.

Gell-Mann: No, the SU(2) group not only has a doublet representation, but also a triplet representation. Similarly, the SU(3) group not only has a triplet representation, but also an octet representation. The nucleons and the hyperons form such an octet.

Newton: Since there is a triplet representation, I should see those three particles, but I don't. Where are they?

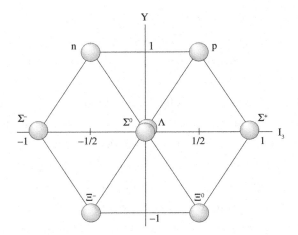

Fig. 3.3. *The octet of the baryons*

Gell-Mann: Well, those triplet objects do not seem to exist, nor do sextet objects. Let me also mention that the hyperons form three representations of the isospin group SU(2): the Lambda hyperon forms an SU(2) singlet; the Sigma hyperons form an SU(2) triplet; and the Xi hyperons form two SU(2) doublets. Thus, the octet of SU(3) consists of a singlet, two doublets, and one triplet.

In addition, the mesons are also described by an octet. The three pions form an SU(2) triplet; the four K mesons form two SU(2) doublets; and the eta[a] meson forms an SU(2) singlet. For the mesons, the symmetry breaking is very large: the pions have masses of about 140 MeV; the K mesons have masses of about 496 MeV; and the eta meson has a mass of about 547 MeV.

Einstein: Thus, the ratio of the mass of a K meson to the mass of a pion is more than three. Nevertheless, the K mesons and the pions are in one representation of SU(3). If the SU(3) symmetry were exact, these masses

---

[a] It is spelled in lowercase because its corresponding symbol is the lowercase greek letter η, in contrast to Lambda (Λ), Sigma (Σ), and Xi (Ξ).

would be the same. Since these masses differ a lot, the symmetry breaking must be very large; for the mesons, it is more than 20%.

Gell-Mann: Yes, the symmetry breaking for the mesons is larger, but this is because the pions have rather small masses.

The SU(3) symmetry should also describe the new particles discovered in the early 1950s at Berkeley. Over there, atomic nuclei were bombarded with pions, and four new, very short-lived particles called Delta resonances were discovered. They have masses of about 1230 MeV and a spin of $\frac{3}{2}$. One of them has an electric charge of +2. These particles are excitations of the nucleons. Later, it was discovered that there are also excitations of Lambda, Sigma, and Xi particles. Thus, there are nine such particles. How could they be described by an SU(3) representation? I concluded that these excitations must form a decuplet, which contains 10 particles. Thus, one particle was missing. This missing particle, which I called "Omega minus", should have an electric charge of −1. I was also able to predict its mass: it is about 1670 MeV.

The search for this particle then began, and in 1964, Nicholas Samios and his group found it in Brookhaven. This discovery convinced most physicists that the SU(3) symmetry is relevant to our Universe. In 1969, I received a Nobel Prize for my work on the SU(3) symmetry.

Newton: You mentioned that the triplet of SU(2) is analogous to the octet of SU(3), and that the doublet of SU(2) is analogous to the triplet of SU(3). Why are there no particles described by a triplet?

Gell-Mann: Good question. In the beginning, I did not know. The observed particles are described by singlets, octets, and decuplets. Triplets and sextets do not exist. But we shall soon see that triplets are relevant: the constituents of the nucleons are triplets.

I propose that tomorrow we drive to Santa Barbara — it will take about two hours — and visit Prof. David Gross, the director of the Kavli institute at the University of California, Santa Barbara (UCSB).

QUARKS

4

The next morning, the four physicists traveled from Pasadena to Santa Barbara. Haller drove on the Ventura Freeway to the west. After they had passed Glendale, Einstein started the discussion.

Einstein: Murray, your SU(3) symmetry is rather abstract in comparison to the isospin symmetry. The nucleons — the proton and the neutron — are a doublet of the isospin group SU(2), and they generate the group. In other words, the doublet is the fundamental representation of the SU(2) group. This is a very natural symmetry, easy to understand. To me, the SU(3) symmetry is, however, rather obscure. The fundamental representation of the SU(3) group is a triplet. Yet, in nature, there are no SU(3) triplets or sextets, only singlets, octets, and decuplets — I cannot understand this. Is there a rule that tells us which representations are allowed and which are not?

Gell-Mann: Well, there is no rule, but now I know why there are no triplets or sextets. In 1964, I studied this problem in detail and introduced the SU(3) triplet particles. They are not normal hadrons,[a] but are the constituents of hadrons: I considered nucleons as bound states of three triplet particles. Also, note that the product of three triplets contains one singlet, two octets, and one decuplet:

$$3 \times 3 \times 3 = 1 + 8 + 8 + 10$$

The eight baryons form an octet, and their electric charges can be used to determine the electric charges of the triplet particles. To my surprise, the charges of the triplet particles are not integers; they are either $\frac{2}{3}$ or $-\frac{1}{3}$. This can be seen quite easily. Let me call the triplet particles "up" (u), "down" (d), and "strange" (s). A proton has the internal structure "uud"; a neutron, "udd". The electric charge of the proton is +1; that of the neutron is zero. Therefore, the electric charges (Q) of the "up" quark and the "down" quark are as follows:

$$Q_u = \frac{2}{3}, \quad Q_d = -\frac{1}{3}$$

Since the charges are not integers, I came to the conclusion that such a model would not make sense.

Einstein: Indeed. And perhaps, the SU(3) symmetry is also wrong.

Gell-Mann: No, the SU(3) symmetry is certainly correct. I kept thinking about the problem with the electric charges. Then, I went to a colloquium at Columbia University in New York. But before the colloquium, I had lunch with my friend Robert Serber, who was a physics professor there. We were conversing about the triplets and their funny electric charges.

---

[a] A hadron is any particle that is made of quarks and interacts via the strong interaction.

Serber remarked:

> Why do you worry about their electric charges? There would be no problem if these triplets are not real particles, but are constituents of hadrons and permanently confined. The forces between the constituents might be so strong that they cannot be liberated.

I had also thought about this possibility, but had not taken it seriously. But after the discussion with Serber, I changed my mind. Moreover, the triplet objects could just be some mathematical symbols, not real particles.

When I returned to Pasadena, I wrote a short paper about the triplet particles. Meanwhile, I had also found a good name for them: "quarks". James Joyce introduced this strange name in his novel "Finnegans Wake", where he wrote: "Three quarks for Muster Mark!" I found this sentence on page 383; since 3 and 8 are important numbers for SU(3), I decided to call the triplet particles "quarks".

Joyce probably picked up this word when he was visiting the city of Freiburg in Germany. At the city market there, one could buy "quark", which in German means a sort of sour cream.

Haller: The word "quark" in German also means "nonsense". Perhaps, this is also suitable for the triplet objects, which cannot exist as free particles: they are "nonsense particles".

Gell-Mann: I didn't know that "quark" also means "nonsense"; otherwise, I would have chosen a different name. When I finished my short paper, I decided to publish it — not in "Physical Review Letters" (PRL), but in "Physics Letters". I assumed that the referees of PRL would have rejected my paper due to the funny electric charges of quarks, if I had sent them my paper.

Einstein: That was a good decision: I would have rejected your paper if I had been one of the referees.

Gell-Mann: The referees of "Physics Letters" also thought that my paper was wrong, but they accepted it because I was a recognized physicist.

Feynman's Ph.D. student George Zweig, a visitor at the European Organization for Nuclear Research (CERN)[b] then, also thought about those quarks and wrote a long article about them, calling them "aces". Because he was at CERN, he had to get at least one of his articles published in a European journal, but Zweig refused to do so, and his paper on quarks was never published, except as a CERN preprint. Later, I arranged for Zweig to be hired by Caltech, and shortly afterwards, he returned to Caltech as a professor.

Newton: You said that quarks are the constituents of baryons and mesons. How do you build the nucleons, for instance, using quarks? There should be specific wave functions that are similar to those of atoms.

Gell-Mann: This is simple. Recall that the three types of quarks are "up"(u), "down"(d), and "strange"(s). Protons and neutrons consist of only u and d: a proton has the internal structure "uud", while a neutron, "udd". The isospin symmetry follows since u and d can be interchanged. Note that, the generators of the isospin group are u and d, not protons and neutrons.

Whilst protons and neutrons consist of only u and d, the other particles in the baryon octet each contain at least one s (see Fig. 4.1). The Lambda particle consists of all three quarks. Its wave function is "uds". The three Sigma particles have the wave functions "uus", "uds", and "dds", and the two Xi hyperons have the wave functions "uss" and "dss".

Newton: The isospin symmetry follows since u and d can be interchanged. But I could also interchange u with s or interchange d with s. This would give two more isospin-like symmetries; the SU(3) symmetry would, then, contain three different SU(2) symmetries. Are these symmetries useful?

---

[b]CERN stands for *Conseil Européen pour la Recherche Nucléaire* (French).

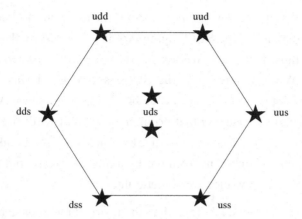

Fig. 4.1. *The eight baryons: Each of them is composed of three quarks.*

Gell-Mann: The isospin symmetry is a good symmetry since the difference in the masses of u and d is very small and can be neglected to a good approximation. But the mass of s is much larger than those of u and d; thus, the symmetries resulting from both of the interchanges (of u with s and of d with s) are strongly broken — broken as strongly as the SU(3) symmetry. Thus, the other two SU(2) groups, which are called U spin and V spin, are not good symmetries compared with the isospin.

Einstein: You told us that quarks are permanently confined and could just be mathematical objects. But now you speak about the masses of quarks. If quarks cannot exist as free particles, they cannot have a mass. A mathematical symbol cannot have a mass.

Gell-Mann: Yes, I introduced the term "mass" for quarks. We shall see later that it is useful to give masses to quarks, especially in the correct field theory of the strong interaction, the theory of quantum chromodynamics, which we shall discuss later.

The SU(3) symmetry would be perfect if the three quarks had the same mass. The breaking of the SU(3) symmetry is generated by the masses of the quarks; thus, the quark masses are some sort of symmetry-breaking parameters. When I introduced the SU(3) symmetry, I was wondering what kind of interaction is needed to break this symmetry. After introducing the quarks, I realized that no interaction is needed: the symmetry is just broken by the masses of the quarks. Therefore, the breaking of the SU(3) symmetry is different from the breaking of the isospin symmetry, which is broken by the electromagnetic interaction.

Newton: You mentioned before that there are three pions, but four K mesons. Can this be understood with the quark model?

Gell-Mann: Yes, this is easy. The three pions are bound states of a quark that is either "up" or "down" and an antiquark that is, again, either "up" or "down".

$$\pi^+ : (\bar{d}u) \quad \pi^- : (\bar{u}d) \quad \pi^0 : \frac{1}{\sqrt{2}}(\bar{u}u - \bar{d}d)$$

The four K mesons are bound states of a quark and an antiquark such that either the quark or the antiquark is "strange".

$$K^+ : (\bar{s}u) \quad K^- : (\bar{u}s) \quad \bar{K}^0 : (\bar{d}s) \quad K^0 : (\bar{s}d)$$

Einstein: I have a question about the SU(3) symmetry. There are two particles with the quark composition "uds": the Lambda particle and the neutral Sigma particle. I would assume that the two particles have the same mass, but they don't. Why?

Gell-Mann: The answer is simple. The masses of the hyperons also depend on the spins of the quarks. In the Lambda hyperon, the spins of u and d are opposite to each other, but in the Sigma hyperon, the two spins are parallel. The mass of a hyperon also depends on the spin structure. One can also understand why the mass of the Lambda hyperon is lower than the mass of the neutral Sigma hyperon, but these details should not be discussed now.

Now, let us consider the Xi hyperons. They have two strange quarks, and their masses are larger than the masses of the Sigma hyperons. It seems that the masses of the hyperons depend strongly on the number of strange quarks. We shall see later that the strange quark has a mass that is about 150 MeV larger than the masses of the up and down quarks.

The masses of the up and down quarks are not equal: the mass of the down quark is slightly larger than the mass of the up quark. Thus, the isospin symmetry is not only broken by the electromagnetic interaction, but also by the quark masses.

Einstein: This is interesting. Since the mass of the down quark is larger than the mass of the up quark, this might explain why the neutron mass is larger than the proton mass.

Gell-Mann: Yes, it does. From the neutron–proton mass difference, one can subtract the mass term due to the electromagnetic interaction, and one finds that the difference between the down quark's mass and the up quark's mass is about 2.5 MeV.

This is the case, fortunately. If the masses of the up and down quarks were the same, the proton mass would be larger than the neutron mass due to the Coulomb field surrounding the proton. As a result, a proton would decay into a neutron. Hydrogen atoms would not exist, and we would not exist since there is a large number of hydrogen atoms in our bodies. In reality, however, a neutron decays into a proton by the weak interaction, and this is due to the fact that the down quark's mass is larger than the up quark's mass.

Now, let us consider the decuplet (see Fig. 4.2). The lightest particles in the decuplet are the four Delta particles with masses of about 1230 MeV. The Delta particle with an electric charge of +2 consists of three up quarks: its wave function is "uuu"; the spins of all of its quarks are parallel. The other

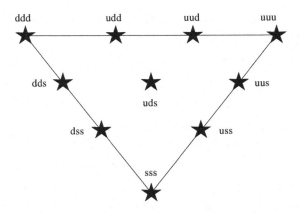

Fig. 4.2. *The baryon decuplet and its quark composition*

three Delta particles have the wave functions "uud", "udd", and "ddd'. The Sigma resonances are heavier; their masses are about 1385 MeV, and their wave functions are "uus", "uds", and "dds". The Xi resonances' masses are about 1530 MeV, and their wave functions are "uss" and "dss". The heaviest particle in the decuplet is the Omega minus, the king of the decuplet, which has a mass of about 1672 MeV. It consists of three strange quarks.

Einstein: Murray, look, we've just arrived at the beach. I remember from previous visits that the freeway from here to Santa Barbara runs alongside the beach. After so many years, I finally see the Pacific Ocean again!

Soon, their car entered Santa Barbara. After passing the city center, Haller left the freeway and took Route 217 to UCSB, which is just behind the airport. A few minutes later, they arrived at the Kavli Institute for Theoretical Physics (KITP), which is near the university entrance.

Gell-Mann and Haller went to the office of David Gross, the director of KITP, who put a large office at their disposal. The three then went for lunch in a cafeteria close to the Pacific Ocean, before taking a long walk along the beach.

COLORED QUARKS

5

After lunch, Einstein, Gell-Mann, Haller, and Newton went to their office at the Kavli institute. The discussion started after coffee.

Gell-Mann: Now, I would like to mention a fundamental problem of the quark model. Let us study the Omega-minus particle in more detail. In the quark model, it is a bound state of three strange quarks. The three spins of the three quarks are aligned; thus, the spin of the Omega minus is given by those of the three quarks. If quarks are fermions, like electrons, there is a problem with the Pauli principle. As you know, the wave function of a state must change its sign if two fermions are exchanged. However, the wave function of the Omega minus ("sss") is symmetric if two quarks are exchanged.

Einstein: You did not consider the space wave function. It might be antisymmetric; the space wave function could be a p-wave.

Haller: Dear Albert, the Omega minus is in the ground state of three strange quarks, and we know from quantum mechanics that a ground-state wave function is always symmetric; it cannot be a p-wave.

Einstein: Sorry, I have forgotten this rule of quantum mechanics. Then, I see a serious problem here. Murray, your quark model is probably wrong.

Gell-Mann: Many physicists, indeed, thought that the quark model is nonsense, due to this statistics problem. I also worried about it, and in my first paper about quarks, I did not mention it. But shortly after the publication of my letter on quarks, many physicists considered possible solutions.

Oscar Greenberg, a professor at the University of Maryland, proposed that quarks are not normal fermions, but have their own statistics, called parastatistics. But I did not believe that quarks have a funny statistics; I did not even know what parastatistics is.

My friend Yoichiro Nambu in Chicago and his collaborator Moo-Young Han investigated another possibility. They introduced a new type of SU(3) symmetry, and they were able to have integrally charged quarks. Thus, those quarks could be seen in experiments as real particles.

In the Han–Nambu scheme, there are three different up quarks: two with the charge +1 and one with the charge 0. Instead of three quarks, there are nine quarks in total. If the charges of the three different quarks (+1, +1, 0) are averaged, one obtains $\frac{2}{3}$, the charge of the up quark in my quark model. I did not like the integer charges in this model, but we shall soon see that Han and Nambu made a step in the right direction.

In 1970, I started to collaborate with the young German physicist Harald Fritzsch. He was then a Ph.D. student at the Max Planck Institute in Munich. I met him in the summer of 1970 in Aspen and invited him to Caltech. In the fall of 1970, Fritzsch moved to the Stanford Linear Accelerator Center (SLAC) at Stanford University in Palo Alto for six months. He came regularly to Caltech, about twice a month and for a few days each time. He liked to drive his big old Dodge car; he always drove from SLAC to Caltech and back on the coastal Highway 1, always stopping in the area of Big Sur for a night.

Fritzsch and I considered the problem of statistics again; we studied in particular the Han–Nambu scheme and modified this scheme in the sense that the additional group SU(3) is assumed to be an exact group. In our modified scheme, the electric charges of the quarks are, as usual, $\frac{2}{3}$ and $-\frac{1}{3}$, but besides spin and electric charge, the quarks have an additional property, which we called "color".

Let us consider an up quark. The quantum number "color" implies that this quark can appear as a red quark, a green quark, or a blue quark. The color symmetry is described by the color group SU(3).

With color, one has new possibilities to construct the wave function of a baryon. Let us consider three red up quarks. Again, there is the same problem of statistics, since the wave function is symmetric if two quarks are exchanged. Now, let us consider the antisymmetric combination:

red/green/blue − red/blue/green + blue/red/green − blue/green/red + green/blue/red − green/red/blue

For a baryon with this type of wave function, there is no problem with the statistics. If two quarks are interchanged, the wave function changes its sign — as requested by the Pauli principle — since the color wave function is antisymmetric.

Haller: I'd like to emphasize that the antisymmetric sum that you just mentioned is a singlet under the color group SU(3). Thus, a baryon is a color singlet. There is also another possibility to construct a singlet if we take the quarks together with the antiquarks:

anti-red/red + anti-green/green + anti-blue/blue

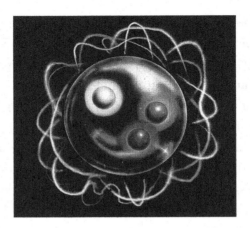

Fig. 5.1. *A proton consists of three colored quarks.*

This sum is also a color singlet, and it describes the mesons. We have a simple rule: all observed hadrons (such as protons, see Fig. 5.1) are color singlets.

Gell-Mann: Yes, the simplest color singlets are the three-quark states (the baryons) and the antiquark–quark states (the mesons). If we require that hadrons be color singlets, then there would be no hadrons consisting of two quarks since those would either be sextets or anti-triplets.

Newton: The color idea is interesting. Do you know any additional facts that indicate that quarks have this additional degree of freedom (color)?

Gell-Mann: Yes, Fritzsch and I discovered one of them in the fall of 1971 at CERN, where we were both staying for a year. Together with William Bardeen, we investigated the electromagnetic decay of the neutral pion, which decays very fast into two photons. Jacques Steinberger, now a famous experimentalist at CERN, was a theorist when he was young. Around 1950, he calculated the decay rate of the neutral pion by assuming that the neutral pion is a nucleon–antinucleon bound state. He obtained a result that agrees perfectly with experimental observations.

It was known that the observed decay rate does not agree with that predicted by the quark model — the predicted rate is smaller than the observed rate by about a factor of nine — due to the non-integer charges of the quarks. But we found out that the strength of the pion decay depends on the number of colors. If there are no colors, the rate is smaller by a factor of nine; if there are three colors, the decay amplitude is three times larger because each color contributes to the amplitude, and the decay rate is nine times larger (than that predicted by the original model with no colors). Thus, the model with three colors agrees perfectly with experimental observations. This was another sign for the existence of colors. I had the feeling that we were on the right track.

Something else is important in this connection. Fritzsch had worked in Potsdam, East Germany, on a diploma thesis on gravity. His professor told him that he should first study the Yang–Mills theory.

Einstein: I am an expert on gravity, but I do not know what Yang–Mills theory is.

Haller: In 1954, the Chinese physicist Chen Ning Yang and his collaborator Robert Mills worked at the Institute for Advanced Study in Princeton. They were interested in the strong interaction and tried to give a fundamental meaning to the isospin symmetry group. They had the idea of using the isospin symmetry in the same way as the gauge symmetry in electrodynamics. The gauge symmetry of electrodynamics is given by the gauge group $U(1)$, which describes the phase transformations of an electron field; these transformations depend on space and time. A phase rotation of an electron field must be accompanied by a change in the potential that describes the photon. This is the gauge theory of electrodynamics, in which electron and photon fields are closely linked.

Yang and Mills used the isospin group $SU(2)$ instead of the $U(1)$ group as a gauge group. The $SU(2)$ group describes the phase rotations of a nucleon field; these rotations also depend on space and time.

Einstein: But then, there should also be gauge particles analogous to the photons in electrodynamics. Where are they?

Haller: Since the gauge group is SU(2), there are three gauge bosons. They would be massless like photons, but we know that such particles do not exist. Yang and Mills argued that these gauge bosons might have a mass and could be discovered soon by experiment. But they could not find a way to introduce mass to the gauge bosons; thus, their theory was not realistic.

I should also mention that such a theory had been constructed in 1953 by Wolfgang Pauli. He wrote a long letter to Abraham Pais, who was at the Institute for Advanced Study in Princeton. Pauli also faced the problem of introducing mass to the gauge bosons; for this reason, he did not publish his theory. Pauli's letter had also been read by Yang and Mills, but they did not mention the letter in their paper. Regardless of what they did or did not do, "Yang–Mills" theory should be called "Pauli–Yang–Mills" theory.

Gell-Mann: Fritzsch knew much about Yang–Mills theory, and one day, he proposed to use the color group as a gauge group. At first, I did not like the idea, but after one day, I thought that it might be right. Unlike Yang and Mills, we were able to accommodate the massless gauge bosons since they are colored and, therefore, confined like quarks.

I proposed to call the gauge particles "gluons" (see Fig. 5.2), a name derived from the word "glue". Fritzsch did not like it and proposed the name "chromons" for the gauge particles. I wish we had chosen this name, because the name "gluon" is rather funny: it is a combination of Greek and English. Nevertheless, the name "gluon" was then generally accepted. But Fritzsch and I found a good name for this gauge theory: we called it quantum chromodynamics (QCD).

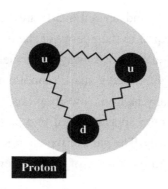

Fig 5.2. *The three quarks in a proton are bound by gluons.*

Einstein: Yes, this is a very nice name. It sounds better than quantum electrodynamics (QED), but is the theory really correct?

Gell-Mann: It seems so, but in the beginning, many physicists did not like it. However, in 1974, things changed. By then, our theory was very popular. Fritzsch obtained several offers for full professorship in Germany. In 1977, he accepted one of those offers and left Caltech, but he returned every year, and we continued our collaboration. In 1980, Fritzsch went back to Munich as a professor at the Ludwig Maximilian University of Munich, the leading university in Germany.

Einstein: I like your theory on the strong interaction, especially the fact that the color symmetry is an exact symmetry. A symmetry that is not broken is remarkable; this is not often the case in our world. God has introduced the colors, but in such a way that no one can see them directly. It took quite a while before Fritzsch and you found the colors of quarks.

Gell-Mann: The experimental physicists also found hints that gluons exist and that the color symmetry is relevant to quarks. In 1975, Richard Feynman pointed out that one could see quarks indirectly during the annihilation of an electron and a positron at very high energies. During the annihilation, a quark and an antiquark are created. Both have the same energy, and it is equal to the energy of the electron or the positron. The quark and the antiquark move away from their point of creation at virtually the speed of light. Since a quark cannot come out as a free particle, it fragments into many hadrons, mostly pions. (The same goes for an antiquark.) The result is the appearance of two particle jets, which could be observed experimentally. The sum of the energies of all the hadrons in one jet should be equal to the energy of the electron or the positron.

Haller: Feynman explained his idea of jets to me in 1975, but I did not take it seriously. But in 1978, the two jets were clearly seen at the German Electron Synchrotron (DESY)[a], Hamburg (see Fig. 5.3). Feynman was right: quarks can be seen indirectly. Later, gluons were also seen in the same way. Murray and I had already found in 1970 that gluons contribute about 55% of the momentum of a fast-moving nucleon, whereas quarks contribute only 45%.

According to QCD, a gluon should also — and often — be emitted in the creation of a quark and an antiquark. Since gluons are colored, they would also fragment into hadrons. Thus, in an electron–positron annihilation, one should sometimes see three particle jets. In 1979, the three jets were observed — also at DESY (see Fig. 5.4). Two of the jets come from the quarks, and the third jet is due to the fragmentation of the gluons.

Now, I would like to point out an important difference between QCD and QED. The gauge symmetry of QED is rather simple: it is the symmetry of phase rotations. We can rotate the complex plane by 30 degrees forward, then, 10 degrees backward, etc. A rotation can always be described by a parameter: the corresponding angle. If two rotations are carried out, say by 10 degrees and then by 30 degrees, the result is the same as a rotation by 30 degrees and then by 10 degrees. The final result does not depend on the order of the two rotations. Mathematicians call

---

[a] DESY stands for *Deutsches Electronen-Synchrotron* (German).

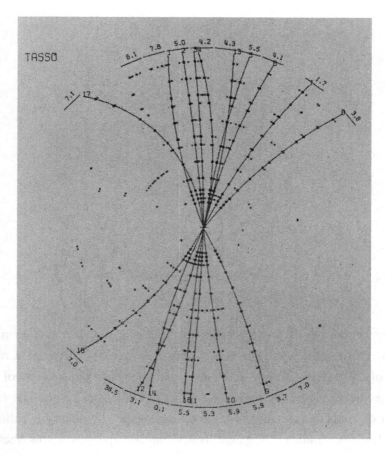

Fig. 5.3. *Two quark jets, first observed in 1978 with the TASSO detector at DESY*

this an Abelian symmetry, named after the Norwegian mathematician Nils Henrik Abel, who lived in the 19th century.

The symmetry of the colors, however, is more complicated. Let me describe this by using a geometric example. A rotation in a two-dimensional space is described by just one angle. But in a three-dimensional space, there are three different rotations: one about the *x*-axis, one about the *y*-axis, and one about the *z*-axis. A rotation in a three-dimensional space is described by three parameters. Suppose we

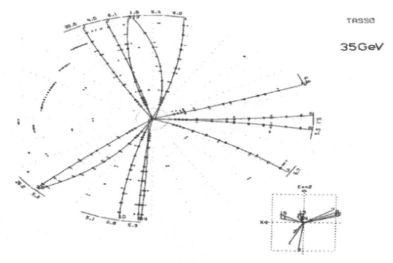

Fig. 5.4. *A three-jet event, first observed in 1979 at DESY*

arrange a rotation of 10 degrees about the *x*-axis and then another rotation of 20 degrees about the new *y*-axis. If we, instead, arrange a rotation of 20 degrees around the *y*-axis and then another rotation of 10 degrees around the new *x*-axis, we will get a different result. The final result depends on the order of the rotations. Such a symmetry is called a non-Abelian symmetry. A gauge theory that incorporates a non-Abelian symmetry is called a non-Abelian gauge theory. The theory that was first discussed by Pauli and then by Yang and Mills is a non-Abelian gauge theory.

Yang and Mills thought that they could describe the strong interaction with a non-Abelian gauge theory. However, they could not solve the problem of introducing mass to the gauge bosons; a way of doing so was only found around 10 years later. We shall discuss this later when we discuss the weak interaction in more detail.

Einstein: You mentioned that Fritzsch worked in Potsdam, in the former German Democratic Republic. As we know, citizens of that country could not travel to the West. He was in Potsdam and then in Pasadena. How could this happen?

Gell-Mann: As a student in East Germany, Fritzsch also worked in politics. He arranged actions against the communist regime, which are still well known in Germany. If the secret police had caught him, he would have been condemned to at least 15 years in prison. He escaped in the summer of 1968 with a friend to West Germany, crossing the Black Sea in a small boat from Varna in Bulgaria to Turkey — 300 km with just a small outboard engine! This escape impressed me, and I started to collaborate with Fritzsch.

He and I were working out the details of the color theory. In September 1972, the Rochester conference on particle physics was held in Chicago. Both Fritzsch and I were there, and I gave a talk on our gauge theory. The name "quantum chromodynamics" was introduced one year later. In 1973, more details of the theory were found and published by us, together with Heinrich Leutwyler from the University of Bern.

Now, I would like to mention a few details about this theory. Its gauge group is the color group, that is, the SU(3) group. Its gauge bosons form an octet of this SU(3) group: there are eight different gluons. This can be seen as follows. A quark carries one of the colors "red", "green", or "blue". If a quark interacts with a gluon, the color of the quark is changed to a different one. There are nine different transitions:

$$\text{red} \to \text{red}, \text{red} \to \text{blue}, \text{red} \to \text{green},$$
$$\text{blue} \to \text{red}, \text{blue} \to \text{blue}, \text{blue} \to \text{green},$$
$$\text{green} \to \text{red}, \text{green} \to \text{blue}, \text{green} \to \text{green}.$$

One of them does not count because the sum

$$(\text{red} \to \text{red}) + (\text{blue} \to \text{blue}) + (\text{green} \to \text{green})$$

is a singlet of the color group. Thus, there are only eight possibilities, which correspond to the eight different gluons.

QCD is very similar to QED. In QED, photons transmit electro-magnetic forces between electrically charged particles; in QCD, gluons transmit chromodynamic forces between colored quarks and gluons.

The photons in QED are neutral; there is no direct electromagnetic interaction between them. In a laser beam, many photons fly together through space without any interactions.

In QCD, things are different. Gluons are colored; they interact not only with quarks, but also with other gluons. Thus, in QCD, the vacuum polarization (of the gauge boson) is different from that in QED. Let us consider a single quark surrounded by virtual quark–antiquark pairs. The color charge of the quark pushes the virtual quarks away, but attracts the virtual antiquarks. This effect is similar to the related effect in QED, where the electric charge of the quark pushes the virtual quarks away and attracts the virtual antiquarks. It follows that the color charge of a quark is partially screened.

However, due to the self-interaction of gluons, the sea of the virtual gluons is also affected. It turns out that this effect is different: it does not lead to a screening of the color charge; it increases the color charge. The initial calculations in this direction were done by Iosif Kriplovich in the Soviet Union and by Gerard 't Hooft in the Netherlands.

Haller: David Gross was also interested in the vacuum polarization of gluons. He had a good Ph.D. student called Frank Wilczek, who started to work on this problem. Likewise, at Harvard University, Sidney Coleman's Ph.D. student David Politzer also calculated the details of the vacuum polarization of gluons in QCD.

Gross, Wilczek, and Politzer found in 1973 that virtual gluons, indeed, increase the color charge; thus, in QCD, the vacuum polariza-tion is different from that in QED.

The coupling constant of gluons to quarks increases as energy decreases, and it decreases as energy increases (see Fig. 5.5). The QCD coupling constant at a given energy is described by a constant denoted

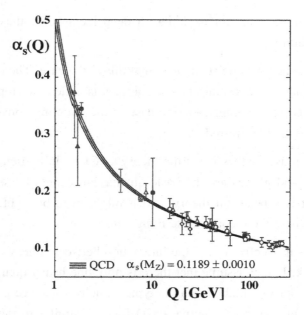

Fig. 5.5. *The QCD coupling constant as a function of energy*

as α_s, which is similar to the fine structure constant, α, in QED. The empirical value of the QCD coupling constant, α_s, at the energy equivalent to the mass of a Z boson,[b] which is about 91 GeV, is as follows:

$$\alpha_s \approx 0.12$$

This coupling constant is rather small, and one can use perturbation theory to calculate the behavior of quarks and explain why quarks can be seen in deep inelastic scattering.

However, the coupling constant increases as energy decreases. Then, the interaction becomes stronger, and perturbation theory no longer works.

Newton: This might explain why quarks do not exist as free particles. They are permanently bound by the QCD interaction and can only be

---

[b]The Z boson is the neutral mediator of the weak interaction.

seen at high energies, for example, in deep inelastic scattering of leptons[c] and hadrons.

Haller: Yes, most physicists believe that this is the case. The decrease in the QCD coupling constant at high energies is called "asymptotic freedom", and correspondingly, the increase in the coupling constant at low energies is called "infra-red slavery".

Einstein: I must confess I like the theory of QCD, in particular the fact that quarks and gluons are the basic quanta, but cannot be seen as free particles. It is a beautiful theory, and it might even be right. Nature is beautiful, so are the theories describing nature.

Haller: It is instructive to see the interaction between a heavy quark and its antiquark. I shall assume that there is only one heavy quark in QCD. At small distances, smaller than the typical nuclear distance given by the radius of a proton, the attractive QCD force is similar to the Coulomb force of electromagnetism: it decreases with the reciprocal of the square of the distance. The QCD force can be described by field lines in the same way as the electric force. But at a distance larger than about one Fermi, QCD field lines attract one another since gluons interact with one another. While electric field lines spread out in space at large distances, QCD field lines at large distances become parallel, in the same way as the electric field lines in a plate capacitor. Thus, the force between a quark and an antiquark does not decrease anymore: it becomes constant. This implies that quarks cannot exist as free particles.

Gell-Mann: Also, the contribution of QCD to our understanding of nuclear forces is interesting. Nuclear forces are responsible for binding six protons and six neutrons into a carbon nucleus. But what exactly are nuclear forces? They are the indirect consequences of the gluonic forces between quarks.

---

[c]A lepton is any elementary, spin-$\frac{1}{2}$ particle that does not interact via the strong interaction.

In atomic physics, there is a similar phenomenon. Two neutral hydrogen atoms and one neutral oxygen atom attract one another (because the atoms consist of electrically charged particles) to form a water molecule. Similarly, there are attractive QCD forces between nucleons because they consist of chromatically charged quarks and gluons. Nuclear forces are nothing but the QCD analogue of the molecular forces in QED.

Haller: To close this discussion, let us consider the various color-singlet bound states that are expected to exist in QCD. The simplest bound states are the quark–antiquark systems, which are the mesons. The lightest meson is the pion: its mass is about 140 MeV. Then, we have the baryons, each consisting of three quarks, and the antibaryons, each consisting of three antiquarks.

Einstein: The baryons and mesons are bound states of quarks. However, two gluons can also be bound and form a color singlet. These particles would be neutral. Do they exist?

Gell-Mann: Fritzsch and I already introduced these particles, which we called "glue mesons", in 1972. We studied their spectroscopy quite extensively. Experimentalists have not yet observed these glue mesons. This is rather strange since we expect their masses not to be very large, but on the order of 1 GeV. However, glue mesons can mix with neutral quark–antiquark mesons, and it is difficult to disentangle them. We think, for example, that an eta-prime meson, which has a mass of about 960 MeV, is a mixture of a glue meson and a quark–antiquark meson.

One can also construct color singlets using two quarks and two antiquarks. Thus far, such mesons, which are called "tetraquarks", have not been observed. Also, four quarks and one antiquark can form a color singlet. These baryons are called "pentaquarks". Again, such particles have not yet been found experimentally.

Haller: I think it's now time for a coffee break. I suggest we continue in about half an hour's time.

THE ELECTROWEAK INTERACTION

6

After a long coffee break, the discussion resumed in the office at the Kavli institute.

Haller: This afternoon, we shall concentrate on the weak force and the current unified theory of the electroweak interaction. Murray, could you first give us an overview of the history of the weak interaction?

Gell-Mann: The weak interaction was discovered at the beginning of the last century. It was observed that some atomic nuclei are unstable and decay into other nuclei by a process that emits an electron. This process is called a beta decay. After the details of such a decay were investigated — the energy of the emitted electron and the momentum of the nucleus after the decay were measured — it was realized that there was a big problem. The sum of the energy of the nucleus after the decay and that of the electron should be equal to the energy given by the mass of the nucleus at rest before the decay. But the measured energy after the decay was less than expected: energy was not conserved. Also, momentum and angular momentum were not conserved.

Niels Bohr considered the possibility that energy, momentum, and angular momentum are actually conserved but appear otherwise in quantum physics due to the uncertainties in these quantities. But in this case, one would expect the energy to be sometimes larger and sometimes smaller than expected. However, energy is always lost in a beta decay.

Wolfgang Pauli did not believe that energy is not conserved. In 1930, he had the idea that in a beta decay, not only an electron, but also a neutral particle, which could not be observed, is emitted. If the energy, momentum, and angular momentum of this neutral particle are considered, the conservation laws should hold true.

Later, this neutral particle was given the name "neutrino", which means small neutron. Pauli believed that his hypothesis could never be tested, since neutrinos could never be observed. But he was wrong. In 1956, an experiment was carried out in a big nuclear reactor in the Savannah River Site, South Carolina, by Clyde Cowan and Frederick Reines. The neutrinos emitted from the reactor were observed: they collided with protons in a detector, and each of them produced a neutron and a positron; both the neutron and the positron were observed by Cowan and Reines.

Einstein: A beta decay involves the interaction of four particles (namely, the decaying neutron and the three particles produced — a proton, an electron, and a neutrino). How can one describe such an interaction of four particles? In electrodynamics, there is a force particle — the photon. But there is no such particle here.

Haller: In 1934, Enrico Fermi, an Italian physicist, published a theory of beta decay. He posited that the interaction in a beta decay is just an interaction of four fermions and that the strength of this interaction is given by a constant known as the Fermi constant, which has the physical dimension of the inverse of the square of mass. Fermi's theory worked rather well, but we shall soon see that his theory has a lot of problems.

Today, we know that the weak interaction is also mediated by a force particle — the weak boson, which unlike the photon, has a large mass. We shall discuss this later.

Newton: Neutrons and protons are bound states of quarks. How can one describe a beta decay in the quark model?

Gell-Mann: In 1958, Richard Feynman and I published a more elaborate theory of the weak interaction. We postulated that the strength of the weak interaction in a beta decay is given by the product of two currents (one consisting of an electron and a neutrino, and the other consisting of two quarks) and the Fermi constant. In a beta decay, a down quark decays into an up quark, an electron, and an antineutrino.

I must mention another peculiarity of the weak interaction. In 1956, it was discovered that the weak interaction violate the symmetry of parity.

Newton: What is parity?

Gell-Mann: It is the symmetry with respect to space reflection. Let us consider a space with only two dimensions. Then, we can describe every point (denoted as P) in space by two numbers, two coordinates as follows: $P = (x, y)$. A space reflection is given by the transformation $x \to -x$ or $y \to -y$. Such a reflection can also be achieved by a rotation of 180 degrees. Thus, in a two-dimensional space, reflection and rotation are equivalent.

This is, however, not the case in our three-dimensional space. A reflection, that is, the transformation $x \to -x$, $y \to -y$, or $z \to -z$, cannot be obtained by one or more rotations. Rotation and reflections are two different things. It's natural to ponder whether the reflection symmetry is an exact symmetry in nature. Of course, we know that in classical mechanics, the reflection symmetry is preserved: if we consider a mechanical process and its image in a mirror, the process we see in the mirror can also be reproduced in reality.

It was generally assumed that the interactions observed in particle physics also respect the reflection symmetry. Indeed, the reflection symmetry is preserved by the electromagnetic and the strong interactions. But in 1956, it was discovered that the reflection symmetry is violated by the weak interaction. It turns out that the electrons emitted in beta decays are all left-handed. If a left-handed electron is viewed in a mirror, it appears right-handed. Thus, the beta decay seen in a mirror could not be realized in nature: parity is violated.

Einstein: Parity would also be violated even if the electrons emitted in beta decays were partly left-handed and partly right-handed. It is strange that the electrons produced in beta decays are purely left-handed. Why should they be left-handed?

Haller: Yes, this could be the case, but we observe that the violation of parity is maximal. Murray, together with Richard Feynman, published a theory on the maximal violation of parity in 1958. They postulated that all weak currents are purely left-handed. Thus, only left-handed leptons and quarks take part in the weak interaction.

Newton: I have a question on hyperons, which we discussed earlier. How does a Lambda baryon decay?

Gell-Mann: The strange quark in a Lambda baryon decays like the down quark in a neutron, and parity is also violated in the decay of a Lambda baryon. The decay of a Lambda baryon, however, is not similar to the decay of a neutron. Instead, it proceeds only via a "mixing": its weak current is a mixture of down and strange quarks; it contains a down quark multiplied by the cosine of an angle and a strange quark multiplied by the sine of the same angle. Thus, the sum $d \cdot \cos \theta_c + s \cdot \sin \theta_c$ frequently appears in the calculation of the weak interaction. I introduced this angle in 1963; however, it is called the Cabibbo angle, named after the Italian physicist Nicola Cabibbo, who published his theory at the same time as I did.

The measured Cabibbo angle is about 13 degrees. Were it zero, a Lambda baryon would be stable, like a proton.

Einstein: Thus, a weak current connects a u quark with a mixture of d and s quarks. What about the orthogonal sum $-d \cdot \sin\theta_c + s \cdot \cos\theta_c$? Is this sum also associated with the weak interaction?

Haller: Good question, if we assume that this orthogonal combination does not take part in the weak interaction, there will be problems related to neutral currents, which we shall discuss later.

In 1970, at Harvard University, Sheldon Glashow together with two collaborators, proposed that this orthogonal sum interacts weakly with a quark they called the "charm" quark, which, they believed, should also exist. This quark, if it exists, would be a heavy relative of the up quark. According to their scheme, there are two quarks with the charge $-\frac{1}{3}$ and two quarks with the charge $+\frac{2}{3}$. They showed that in this case, the problems with neutral currents disappear.

With the addition of the charm quark, the scheme of the quarks is now similar to the scheme of the leptons. Besides the electron and its neutrino, there is also a heavy brother of the electron — the muon — with a mass of about 105 MeV. This new lepton also has its own neutrino. Thus, there are now four leptons and four quarks; they can be viewed as two lepton doublets and two quark doublets:

$$\begin{pmatrix} v_e \\ e \end{pmatrix} \begin{pmatrix} v_\mu \\ \mu \end{pmatrix} \begin{pmatrix} u \\ d' \end{pmatrix} \begin{pmatrix} c \\ s' \end{pmatrix}$$

where d' and s' are mixtures of quarks as follows:

$$d' = d \cdot \cos\theta_c + s \cdot \sin\theta_c$$
$$s' = -d \cdot \sin\theta_c + s \cdot \cos\theta_c$$

The charm quark was discovered in 1974. Soon after that, a new heavy meson that has a mass of about 3.1 GeV and a rather long lifetime was found at the Brookhaven laboratory on Long Island and at SLAC. This meson is a bound state of a charm quark and its antiquark. Two years later, mesons each consisting of a charm quark and an antiquark of another type were found at SLAC. These mesons, known as D mesons, decayed via the weak interaction, as expected. Their charm quarks mostly transform into strange quarks, sometimes, into down quarks.

The measured decay rates of D mesons are in agreement with theory: the amplitude for the decay of a charm quark in a D meson into a strange quark is proportional to the cosine of the Cabibbo angle, and the amplitude for the decay into a down quark is proportional to the sine of the same angle.

Haller: In 1975, a new charged lepton — the tau lepton — was also discovered at SLAC. Its mass is rather large, about 1780 MeV; thus, the name "lepton" is actually not appropriate for this particle, since "lepton" means "light particle". This lepton also has its own neutrino — the tau neutrino — which decays quickly via the weak interaction.

One year later, in 1976, a new heavy meson — the "Upsilon" meson — was discovered at Fermilab. Its mass is about 9.5 GeV. Soon, it became clear that this meson does not consist of any of the four quarks we had discovered thus far. It is a bound state of a fifth quark, which was named the "bottom" quark (b quark) and has an effective mass of about 4.3 GeV; the Upsilon meson is a bound state of a bottom quark and its antiquark.

Newton: If this is the case, there must also exist mesons each consisting of a bottom quark and one of the other antiquarks — for example, a meson consisting of a bottom quark and an up antiquark — which have masses of about 5 GeV. Is something known about these new mesons?

Haller: Yes, these mesons were found at SLAC and are called B mesons. The lightest B meson has a mass of about 5.3 GeV. Also, there exist baryons each consisting of at least one bottom quark.

Einstein: Glashow introduced the charm quark, since there should be four quarks if there are four leptons. Now that we have six leptons, there should also be six quarks: besides the bottom quark, there should be also a "top" quark (t quark). The bottom and top quarks are related to the tau lepton and its neutrino.

Haller: You are right. There must be such a top quark. It took many years until it was found at the Fermi National Accelerator Laboratory (Fermilab), using the Tevatron collider. The mass of a top quark is extremely large, about 174 GeV, which is about 35 000 times the mass of an up quark or about the same as the mass of a gold atom.

Due to its large mass, the weak decay of a top quark is very fast. It is faster than the time needed to form a meson, such as the meson consisting of a top quark and an up antiquark. Thus, there exist no T mesons.

Newton: I understand that a top quark decays into a bottom quark, but how does a bottom quark decay?

Haller: It decays like a strange quark, via a mixing. We have discussed before the mixing of the down and the strange quarks, given by the Cabibbo angle. Now, we have three quarks of the charge $-\frac{1}{3}$, whose mixing is described by three angles. The mixing, in general, is given by a unitary 3×3 matrix called the CKM matrix. Here, "C" denotes the name Cabibbo, while "K" and "M" denote the names of two Japanese physicists, Kobayashi and Maskawa, respectively. The Cabibbo angle is the largest mixing angle; the two other angles are rather small.

In 1957, Julian Schwinger studied a model in which the weak interaction is mediated by a force particle in a way similar to the way the electromagnetic interaction is mediated by a photon. According to the model, these weak bosons (force particles that mediate the weak interaction) would have an electric charge; thus, there would be a positively charged boson and a negatively charged boson. Their masses would have to be rather large since they had not been observed. Schwinger assumed that the mass of a weak boson is between 2 GeV and 10 GeV.

In 1981, the then existing proton accelerator at CERN was transformed into a proton–antiproton collider. In 1983, the weak bosons were discovered at this new collider. Their mass was about 80.4 GeV.

Newton: The electromagnetic interaction is described by quantum electrodynamics, which is a gauge theory, and the photon is its gauge boson. Is there an analogous theory for the weak interaction, in which the weak bosons are the gauge bosons?

Haller: Yes, we now know that the weak interaction is also described by a gauge theory. Its gauge bosons, that is, the weak bosons, interact with the weak current given by left-handed leptons or quarks. If a positively charged weak boson interacts with an electron, a neutrino is created. The gauge group of the theory is the symmetry group SU(2).

Einstein: You mentioned the two types of charged weak bosons. But if the gauge group is SU(2), there must also be a neutral gauge boson, since the gauge bosons form a triplet of SU(2). Could this neutral particle be the photon?

Gell-Mann: Unfortunately, this is not possible. The neutral boson would interact only with left-handed electrons, but a photon interacts with both left-handed and right-handed electrons.

Einstein: Ok, then a unification of the electromagnetic and the weak interactions is not possible. We must consider a gauge theory that is applicable only to the weak interaction. Then, there must also be a neutral gauge boson with a large mass, and the gauge group would be SU(2). The neutral gauge boson would mediate the neutral weak interaction. If a neutrino interacts with an atomic nucleus, it would remain a neutrino. Has such a neutral-current interaction been observed?

Haller: Yes, the neutral-current interaction was discovered at CERN at the beginning of the 1970s. However, it turns out that the current describing this interaction is not purely left-handed. It also contains right-handed particles.

Einstein: Very strange, the normal weak interaction violates parity maximally, but the neutral-current interaction does not. What kind of gauge group would this imply?

Gell-Mann: It turns out that the simplest theory is one that provides a unification of the electromagnetic and the weak interactions. From 1964 until 1968, Shelley Glashow, Abdus Salam, and Steven Weinberg studied a gauge theory based on the gauge group $SU(2) \times U(1)$. In that theory, the electromagnetic and the weak interactions are unified. Therefore, one speaks of an electroweak theory.

In this unified model, there are two neutral gauge bosons (namely, the Z boson and the photon). Moreover, there are two pure mass eigenstates. You might think that one of them is the photon, but this is not the case. The two neutral bosons are not pure mass eigenstates, but are mixtures of the two pure eigenstates. There is one heavy mixed mass eigenstate; it is the Z boson. Its mass is slightly larger than the mass of the charged weak bosons. The other mixed mass eigenstate is massless; it is the photon. Thus, both the Z boson and the photon are mixed gauge bosons.

Einstein: This is a rather funny model. A photon is a mixture? The massless photon should not be a mixed state. Is nature really so complicated?

Haller: Yes, the experiments of the last 20 years confirm that the model based on the gauge group $SU(2) \times U(1)$ is correct. The mass of a Z boson has been measured to be 91.2 GeV. As expected, this mass is larger than the mass of the charged weak bosons due to the mixing with a photon.

Newton: The weak bosons have big masses, much larger than the mass of a proton, and the photon is massless. The mass of a proton is understood in QCD: it is just the field energy of the quarks and gluons inside the proton. But where do the masses of the weak bosons come from? Can one calculate these masses from first principles? Is the electroweak theory also renormalizable like the theory of quantum electrodynamics?

Haller: Nobody can calculate the masses of the weak bosons. We do not know why the masses of the charged weak bosons are 86 times larger than the mass of a proton. Also, it is not possible to insert these masses into the equations of the SU(2) × U(1) gauge theory. Thus, the theory is not renormalizable, that is, infinities appear in the calculation. However, one can avoid these problems if one assumes that the masses of the weak bosons are generated by a spontaneous breaking of the gauge symmetry.

Einstein: What is a spontaneous breaking of a symmetry?

Haller: I shall give you a simple mechanical example. Let us consider a Mexican-hat potential, that is, a potential that looks like a Mexican hat (see Fig. 6.1). Such a hat is symmetric under rotation: I can rotate the hat about the symmetry axis, which I name it as the z-axis, and you will not notice a difference. Now, suppose I place a small ball at the highest

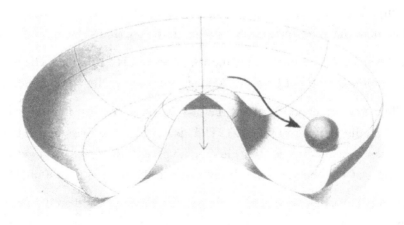

Fig. 6.1. *A Mexican-hat potential*

point of the hat. As long as the ball is at this point, the rotational symmetry is preserved. But after applying a small perturbation, the ball will roll down and come to rest at some point at the brim of the hat. The ball will now be located at a certain distance from the z-axis. The rotation symmetry is, then, broken. This is the spontaneous breaking of the rotational symmetry.

Such a spontaneous symmetry breaking can also happen in a field theory. Let us consider a system of massive scalar fields that interact with one another. When the mass term has a negative sign, the potential of the fields looks like a Mexican hat. In the mechanical example, the ball is a certain distance from the z-axis after rolling down. In the field example, the distance is a certain vacuum expectation value of the field; the scalar field has a certain non-zero value in vacuum.

Einstein: This vacuum expectation value has the dimension of mass. Is there a connection between this value and the masses of the weak bosons?

Gell-Mann: So far, we have considered only scalar fields, without mentioning the weak bosons. But now, we consider a field theory in which we not only have scalar fields but also the weak bosons. The interaction of the scalar fields with the weak bosons generates masses for the weak bosons that are proportional to the vacuum expectation value. If we assume that such a theory describes the weak interaction, then the vacuum expectation value is fixed by the Fermi constant to about 294 GeV, a rather big value, about 300 times larger than the mass of a proton. The mass of the charged weak boson is given as the product of this value and the weak coupling constant, which is not known, but can be determined experimentally.

In 1964, the mechanism of spontaneous symmetry breaking was used to generate a mass for the gauge boson in quantum electrodynamics, that is, a mass for the photon (the mass of the photon is zero). This mechanism for generating a mass for a gauge boson was introduced by several physicists, such as Peter Higgs, a professor in Edinburgh. Thus, this mechanism became known as the "Higgs mechanism".

Since a scalar field is used to generate mass, a scalar boson often called the "Higgs boson" should exist. In 1971, it was shown by Martinus Veltman in Utrecht and his Ph.D. student Gerard 't Hooft that in a non-Abelian gauge theory, no infinity arises if the masses of gauge bosons are generated by the Higgs mechanism.

Haller: We do not know whether the masses of weak bosons are generated by the Higgs mechanism. If this is the case, then there must also exist a scalar Higgs boson. Several searches for this boson have been made at the Large Electron–Positron Collider (LEP) at CERN, but without success. The current lower limit on the mass of a Higgs boson found by the LEP is about 115 GeV.

With the new Large Hadron Collider (LHC) at CERN, one should be able to find the Higgs boson soon. If not, the masses of the weak bosons would most probably be generated in a different way unknown thus far. In 2012, a signal that might have come from the decay of a Higgs boson was seen at the LHC. The mass of this observed boson is about 125 GeV. Further experiments are needed to determine whether this boson is, indeed, the Higgs boson.

Newton: I do not understand why a Higgs mechanism is needed. The mass of a proton is dynamically generated and given by the field energy of its gluons and quarks. The masses of weak bosons should be generated by a similar dynamical mechanism; the Higgs mechanism is rather strange. If the masses of weak bosons are due to the Higgs mechanism, there will be two different ways to generate mass in the Universe. I do not like this.

Einstein: Correct, mass is just a specific form of energy, and therefore, all masses in the Universe should be generated by a dynamical mechanism like the one in QCD, which gives the masses of hadrons. I think the Higgs mechanism is just wrong.

Haller: I agree, and perhaps, the masses of weak bosons are due to a new theory in which weak bosons are bound states of some smaller constituents. A long time ago, I considered such models, in which weak bosons consist of two constituents, which I called "haplons". Gell-Mann had invented this name. It is derived from the Greek word "haplos", which means simple.

If weak bosons are bound states of two haplons, similar to the mesons in QCD (mesons are bound states of two quarks), one would expect that there also exist excited states, that is, excited weak bosons, which would have spins 0, 1, or 2. The boson observed at the LHC might be a spin-0 excited weak boson. We shall soon get to know more about this from the experiments at the LHC.

But now, we should stop our discussion. David Gross has just arrived, and we shall go with him to a restaurant down at the harbor of Santa Barbara.

OSCILLATING NEUTRINOS

7

The next morning, they drove back to Pasadena. After having had lunch in the Athenaeum, they continued the discussion in the Athenaeum's small library.

Haller: Today, I will start by mentioning a strange phenomenon related to the quantum nature of neutrinos. As you know, neutrinos were introduced by Pauli around 1930. Since neutrinos are neutral, they do not experience the electromagnetic interaction, but they take part in the weak interaction.

When Pauli introduced the neutrinos, he assumed that there is only one type of neutrino — the one that is produced in a beta decay. But a new lepton — the muon — was later discovered. A muon is unstable and lives on average only 10^{-6} s after it is produced in a particle collision. Like an electron, a muon is also associated with a neutrino — the muon neutrino.

In 1975, a new charged and unstable lepton — the tauon — was discovered at SLAC. It has a large mass, about 1777 MeV, which is about twice the mass of a proton. This particle should not be called a lepton, since lepton means "light particle". Associated with the tauon is also a new neutrino — the tau neutrino. Thus, in our Universe, there are three charged leptons and three uncharged neutrinos.

Newton: Again, the number "three" seems to play a role: our space has three dimensions, quarks have three colors, and there are three pairs of leptons. Is this an accident?

Gell-Mann: We have no idea why there are three colors and three neutrinos. It seems that nature likes to play with the number "three".

Haller: Neutrinos are unique: they exhibit a quantum-mechanical effect that is significant even in the macroscopic dimension and is connected with the very small masses of neutrinos. I mean the neutrino oscillation, which was first discussed by Bruno Pontecorvo, an Italian physicist, who spent most of his life in the Soviet Union.

Einstein: I thought that neutrinos are massless, but now you mentioned their small masses. How big are their masses?

Haller: We only know that neutrinos have masses that are very small, but we do not know their values quantitatively. We shall discuss this later.

Let us first consider the neutrino oscillation. The interaction between an electron and a charged weak boson produces a neutrino:

$$e^- + W^+ \rightarrow V_e$$

The same is true for a muon and for a tauon:

$$\mu^- + W^+ \rightarrow V_\mu; \quad \tau^- + W^+ \rightarrow V_\tau$$

If neutrinos have masses, then something strange can happen.

Einstein: I know what you mean. In the case of quarks, there is a mixing: if a charged weak boson interacts with an up quark, then a mixture of mass eigenstates — mostly a down quark, but sometimes, a strange quark — is produced. This mixing is described by the Cabibbo angle.

A similar mixing can also happen for leptons. If a charged weak boson interacts with an electron, a mixture of two or three neutrino mass eigenstates is produced. Since the mass of a neutrino is different from zero, it does not travel through space at the speed of light. If the masses of the three neutrinos produced are different, then their propagation through space changes their flavor since the three neutrino mass eigenstates travel at different velocities. Thus, an electron neutrino can become a muon neutrino, and so on.

Haller: Yes, let us assume, for example, that an electron neutrino is a superposition of two mass eigenstates, and that the mixing angle is 45 degrees:

$$V_e = \frac{1}{\sqrt{2}}(V_1 + V_2)$$

$$V_\mu = \frac{1}{\sqrt{2}}(-V_1 + V_2)$$

When an electron neutrino travels through space, it changes its flavor since the two mass eigenstates have different velocities. It becomes a muon neutrino before turning into an electron neutrino again, and so on. Such transitions are called "neutrino oscillation". They can only give information about the differences in mass between the neutrinos; the absolute values of the masses of the neutrinos cannot be deduced from the neutrino oscillation.

Bruno Pontecorvo already considered such neutrino oscillation in 1957; he considered the oscillations of a neutrino and an antineutrino. This is surprising since the mixing of the flavor of quarks was not known at that time. Also, it was not known whether there exist other types of neutrino besides the electron neutrino.

In 1976, I became interested in the neutrino oscillation, and at that time, I gave a colloquium on them at the Laue–Langevin institute[a] in Grenoble. Rudolf Mössbauer[b] was the director of the institute, and he also became interested in the neutrino oscillation. Shortly after my colloquium, he investigated the neutrino radiation emitted from the small research reactor in his institute. He did not find an effect. Afterwards, he, together with Felix Boehm, who was a professor here (at Caltech), investigated the neutrino radiation emitted from the big reactor in Gösgen, Switzerland. Again, they did not find an effect, but they could derive stringent limits on the oscillation.

Today, we understand why they did not find any effect. Although the neutrino oscillation was discovered in 1998, it turns out that the differences in mass between the neutrinos are very small, less than 0.1 eV. Therefore, the oscillation cannot be detected by an investigation of the neutrino radiation from a reactor in the vicinity of the reactor, such as those done by Mössbauer and Boehm.

The oscillation was discovered in 1998 in Japan. Kamiokande, a big detector, was built in a silver mine in the mountains near the village of Kamioka, south of Toyama, in order to detect the decay of a proton. There, physicists investigated the atmospheric neutrinos that come from the upper atmosphere. These muon neutrinos are produced by the decay of the pions created during the collisions of cosmic-ray particles with atomic nuclei. Further information was later obtained with the Sudbury Neutrino Observatory (SNO) detector in Canada (see Fig. 7.1).

Newton: Since there are three neutrinos, we have three different mass eigenstates. What is known about the mixing of neutrinos? What are the values of the mixing angles?

---

[a] Officially known as the *Institut Laue–Langevin* (French)
[b] The German spelling is *Mößbauer.*

Fig. 7.1. *The SNO detector in Canada*

Haller: At present, we know the values of the mixing angles. Let me write the three different neutrinos as superpositions of mass eigenstates, without indicating the experimental errors on these matrix elements:

$$V_e = 0.09\ v_1 - 0.38\ v_2 + 0.17\ v_3$$
$$V_\mu = 0.17\ v_1 + 0.71\ v_2 + 0.69\ v_3$$
$$V_\tau = 0.38\ v_1 - 0.6\ v_2 + 0.71\ v_3$$

The associated mixing angles are large, much larger than the flavor mixing angles for quarks. The largest mixing angle here is the Cabibbo angle, about 13 degrees, and the sine of this angle is about 0.22. The other two mixing angles are much smaller.

There are also three mixing angles for leptons. Two of them are very large; one of them might be 45 degrees, while the other one is about 33 degrees.

By analyzing the neutrino oscillation, the mass differences between the mass eigenstates can be determined. From experiment, the mass difference between the first and the second neutrino mass eigenstates is about 0.01 eV, and that between the second and the third mass eigenstates is about 0.05 eV. These mass differences are very small. Perhaps, the masses of neutrinos are also of the same order.

Newton: What is known about the length of a neutrino oscillation?

Gell-Mann: The neutrinos coming from the upper atmosphere have been investigated at the Kamiokande detector, and the typical oscillation length is found to be on the order of 50 km. It is interesting to see the effect of quantum mechanics at such a large distance. Of course, this just follows from the very small masses of neutrinos.

Haller: The small neutrino masses are surprising. Why are these masses much smaller than the masses of the charged leptons? Perhaps, the lepton masses are generated by the electromagnetic force. Then, neutral neutrinos would not get a mass in the lowest order of approximation. This might explain why their masses are very small. But the masses of quarks and leptons are still a big mystery.

I think we should take a coffee break now.

THE GRAND UNIFICATION

8

After the break, the discussion continued in the library of the Athenaeum.

Einstein: The Standard Model of particle physics is a combination of the theory of quantum chromodynamics and the theory of the electroweak interaction. These two theories are independent of each other. Is there a possibility to unify them? The gauge group of QCD is SU (3); the gauge group of the theory of the electroweak interaction is SU(2) × U(1). These two groups would have to be subgroups of the same gauge group of some unified theory.

Gell-Mann: Many physicists think that this should be possible, but the energy at which the unification takes place must be extremely large, on the order of 10^{15} GeV.

Newton: I do not understand why such a large energy for the unification is necessary. 10^{15} GeV — this is about the energy of a bacterium, according to Einstein's mass–energy relation. Why can't the unification be achieved at moderate energies, say, at energies on the order of 1000 GeV, or even less?

Gell-Mann: Well, this is not possible. This is because of the difference between the coupling constant of the QCD interaction and that of the electroweak interaction. The quark–gluon interaction is described by a coupling strength that decreases at high energies according to a property called asymptotic freedom. At the mass of a Z boson, about 91 GeV, the strength of the quark–gluon interaction is close to 0.12, while the strength of the electromagnetic interaction is about 0.008, a factor of 15 smaller. In a unified theory, both of these coupling strengths should be the same if one neglects coefficients of order one, which are given by group theory. However, it is only at energies on the order of 10^{15} GeV that the quark–gluon coupling strength is reduced to a value comparable to the electroweak coupling strength. Thus, a unification of the different interactions is only possible at these high energies.

Newton: In a unified theory of QCD and the electroweak interaction, one might also be able to understand the family structure and the electric charges of leptons and quarks. The charges of quarks are $\frac{2}{3}$ and $-\frac{1}{3}$, whereas the charges of leptons are 0 and −1. Perhaps, there is a connection between the colors of quarks and their peculiar electric charges. I noticed that the sum of the charges in each quark–lepton family[a] (or generation) is zero. This property points towards a connection between quarks and leptons and towards a unification.

Haller: Correct, in the electroweak theory, the charges of the leptons and quarks are arbitrary. The charge of an up quark could easily be $\frac{2}{5}$ instead of $\frac{2}{3}$. This is possible since in the electroweak theory, there is a free parameter — the electroweak mixing angle. This angle describes the ratio of the two coupling strengths in the electroweak theory, which

---

[a] For example, the first quark–lepton family consists of the three colored up quarks, the three colored down quarks, an electron, and an electron neutrino.

is based on the gauge group SU(2) × U(1). In some models, there is indeed a connection between the number of colors and the charges of quarks: if there were four colors, the charge of a down quark would not be $\frac{1}{3}$, but $\frac{1}{4}$.

Gell-Mann: Let us now consider specific models of the unification of QCD and the electroweak interaction, that is, models of the "Grand Unification". The gauge group of QCD is SU(3); the gauge group of the electroweak theory is SU(2) × U(1). A unified gauge theory must be based on a gauge group that contains these two groups as subgroups. Also, the correct electric charges must be obtained. As a result, the number of possible groups that can be used is very small.

The smallest of such groups is the symmetry group SU(5). A unified theory based on this group was first discussed in 1974 by Sheldon Glashow and his collaborator Howard Georgi. The members of the first quark–lepton family — the up and down quarks, the electron, and its neutrino — are described in this theory by two representations of SU(5): a representation with 5 elements and another representation with 10 elements.

The five members of the 5-representation are the electron, its neutrino, and the three colored down antiquarks. The sum of the electric charges of these five objects is zero. The charge of $\frac{1}{3}$ of a down antiquark is directly related to the colors of quarks. If we had four colors, the charge of a down antiquark would be $\frac{1}{4}$. Thus, the electric charges of quarks and leptons are quantized correctly.

The ten members of the 10-representation are the six colored up and down quarks, the three colored up antiquarks, and the positron. Again, the sum of their electric charges is zero. Thus, a quark–lepton family is described by 15 particles.

Einstein: You mentioned the electron and its antiparticle, that is, the positron, but only the neutrino, not the antineutrino. Why?

Gell-Mann: It is sufficient to consider only the left-handed particles. In the electroweak theory, neutrinos are massless left-handed fermions. There are no left-handed antineutrinos, but there are left-handed electrons and left-handed positrons. This explains why there are only 15 particles in the first family, not 16, because there are no left-handed antineutrinos.

Einstein: In the SU(5) theory, neutrinos are massless, but we now know that neutrinos have a mass. It seems to me that the SU(5) theory cannot be correct since for neutrinos to have a mass, both left-handed and right-handed neutrinos or left-handed antineutrinos must exist.

Gell-Mann: Yes, this is true, and we have to include the left-handed antineutrinos. But we shall very soon see that there is also another reason why the SU(5) theory is not correct.

Haller: In the SU(5) theory, one can calculate the electroweak mixing angle, θ_w, and find that it is 37.8 degrees. But the observed angle is much smaller, about 28 degrees. Furthermore, the strength of the QCD interaction can be calculated:

$$\alpha_s \approx \frac{8}{3}\alpha \approx 0.02$$

But the experimental value for this coupling constant is about 10 times larger. Thus, there seems to be a problem unless the energy of the unification is very high.

Newton: Thus far, you did not mention the energy at which the unification takes place. The QCD coupling constant goes down at high energies, and if the energy is very large, there might not be a problem.

Gell-Mann: Yes, we can calculate the value of the energy at which the QCD coupling strength is comparable to the electroweak coupling strength. The result is about 10^{16} GeV. At this energy, the electroweak mixing angle would be close to 37.8 degrees, and the QCD coupling strength would be about 0.02. But the SU(5) theory still have some problems, which we shall discuss now.

Einstein: Before we discuss these problems, let me ask a specific question. A unification of the strong and the electroweak interaction implies that quarks and leptons appear together in a representation of the unification group, for example, SU(5). You mentioned the 5-representation, which contains the electron and the three colored down antiquarks. I would expect in this case that the proton is not stable, but decay, for example, into a positron and a neutral meson. Could this happen?

Gell-Mann: Yes, the proton could decay. In the SU(5) theory, there are 24 gauge bosons: the photon, the 3 weak bosons, the 8 gluons and 12 other gauge bosons. These 12 additional bosons, called the X bosons, generate new interactions that are rather strange. For example, an X boson can transform an up quark into a positron. Thus, the decay of a proton is possible. For this reason, these X bosons must have very large masses, on the order of the unification energy, which is about 10^{16} GeV.

A possible decay of a proton is the decay into a positron and a neutral pion, which is a quark–antiquark bound state. Then, the neutral pion decays quickly into two photons.

In the SU(5) theory, the lifetime of a proton depends on the masses of X bosons, which are related to the strength of the symmetry breaking. Experiments tell us that the lifetime of a proton is at least 10^{32} years. The best result so far comes from the experiments performed in an old silver mine near the village of Kamioka, which is in the Japanese mountains about 200 km west of Tokyo.

In this mine, a large pool filled with purified water is installed. This water is observed with many photomultipliers, which are used to detect photons, which would be produced during the decay of a proton. Until today, no proton decay has been observed. Hence, we conclude that the lifetime of a proton must be more than 10^{32} years.

Einstein: I guess this limit on the lifetime implies that the masses of X bosons must be at least 10^{16} GeV.

Haller: Yes, and one could think that there is no problem with the SU(5) theory, but this is not true. With the LEP machine at CERN, the coupling strengths of the QCD and the electroweak interactions were determined to great precision. At very high energies, the coupling strengths should converge to a single value; however, it turns out that they come close to each other, but do not converge to a single value. Thus, there is a problem with the SU(5) theory.

In 1975, another theory based on the gauge group SO(10) was proposed by Harald Fritzsch and Peter Minkowski and also by Howard Georgi. As you know, the SO(3) group describes the symmetry of our three-dimensional space. Similarly, the SO(10) group describes the symmetry of a ten-dimensional space. In our space, an arbitrary rotation can be described by three angles called the Euler angles. Therefore, a gauge theory based on the SO(3) group has three gauge bosons. In contrast, a ten-dimensional space requires 45 such angles. Therefore, a gauge theory based on the SO(10) group has 45 gauge bosons.

In the SO(10) theory, the quarks and leptons in a family are described by just one representation, not by two as in the SU(5) theory. It is the 16-representation: its members are the three colored up quarks, the three colored down quarks, the three colored up antiquarks, the three colored down antiquarks, the electron, the positron, the neutrino, and the antineutrino.

Newton: In the SU(5) theory, we have a 5-representation and a 10-representation — 15 particles altogether. Now, you mentioned 16 particles. There is an extra particle. Which one is it?

Haller: In the SU(5) theory, neutrinos are massless left-handed particles, but in the SO(10) theory, they, like electrons, are normal fermions, with left-handed and right-handed components. The additional particle is the left-handed antineutrino, which is not present in the SU(5) theory and the standard electroweak theory. Thus, neutrinos would have a mass in the SO(10) theory. Since we know that the neutrinos have a small mass, the SO(10) theory might be closer to the truth.

The symmetry breaking in the SO(10) theory might be such that the gauge group first breaks down to SU(5) and then to the QCD gauge group and the electroweak gauge group. However, in this case, we would have the same problem, as in the SU(5) theory, with the convergence of the coupling strengths. But in SO(10), there is another way to break the symmetry: the SO(10) group first breaks down to SO(6) × SO(4).

Einstein: That is interesting. The SO(6) group is isomorphic to SU(4), and the SO(4) group is isomorphic to SU(2) × SU(2). Isaac, if a group is isomorphic to another group, it means, essentially, that the two groups are identical. The SU(4) group would contain the QCD gauge group, SU(3), and the SU(2) × SU(2) group would contain the electroweak gauge group, SU(2) × U(1). Thus, with this alternative way of breaking the symmetry, there might be no problem with the convergence of the coupling strengths.

Haller: Yes, but the details are slightly different. The SU(4) group is the group that you would have if there were no difference between quarks and leptons: leptons are something like particles of a fourth color. But this group, SU(4), is not an unbroken group like the QCD group. After the first stage of the symmetry breaking, the SU(4) and SU(2) × SU(2) groups are broken, perhaps, at the same energy scale. The SU(4) group breaks down to the SU(3) × U(1) group, where SU(3) is the gauge group of QCD, and U(1) describes the quantum number $B - L$, that is, "baryon number minus lepton number". The SU(2) × SU(2) × U(1) group breaks down to the electroweak gauge group, SU(2) × U(1).

In the SU(5) theory, there is only one step of symmetry breaking besides the breaking of the electroweak symmetry:

$$SU(5) \rightarrow SU(3) \times SU(2) \times U(1)$$

In the SO(10) theory, there are two steps of symmetry breaking:

$$SO(10) \rightarrow SU(4) \times SU(2) \times SU(2) \rightarrow SU(3) \times SU(2) \times U(1)$$

One can choose the energy scale of the second step to be between the energy scale of the electroweak theory (about 100 GeV) and the energy scale of the Grand Unification (about 10^{16} GeV), to be about 10^{10} GeV, for instance. In this case, the coupling strengths converge. Thus, there is no problem.

Gell-Mann: Yes, the SO(10) theory might be the correct theory to describe the Grand Unification. But gravitation is still a problem: the gravitational force is still unrelated to the strong or the electroweak force. Thus far, it has not been possible to construct a unified theory of the Standard Model and gravity.

Einstein: I think this will never work. In my theory of general relativity, gravity is not a force like the quark–gluon force, but follows from the curvature of space and time. Please do not unify gravity with the forces of nature. In the case of electrodynamics, it is easy to construct a quantum theory — quantum electrodynamics. But there is no theory of quantum gravitodynamics. A quantum theory of gravity would imply a quantization of space and time. I do not know how to build such a theory.

Haller: I agree with you. I also think that space and time should not be quantized. But I should mention that many theoreticians now think that quarks and leptons are not pointlike, but are manifestations of small, one-dimensional objects called "superstrings". If one uses these superstrings, it seems that one might be able to construct a quantum theory of gravity. However, such a theory is consistent only if our three-dimensional space is actually a ten-dimensional space.

Newton: This seems crazy. We know that our space has only three dimensions.

Haller: Yes, but the seven additional dimensions may be curled up, so that they do not manifest themselves on the macroscopic level. These additional dimensions could be relevant to particle physics. We observe three dimensions of space, three families of quarks and leptons, three colors of quarks and one "color" of leptons. By adding these numbers together, we get ten, just like the ten dimensions of the superstring world. It might be that the symmetries of particle physics come from the geometry of a ten-dimensional space. But this is very speculative and might be totally wrong.

But for now, I propose that we leave the Grand Unified Theories aside and go for lunch in the restaurant of the Athenaeum.

9

THE BASIC CONSTANTS OF NATURE

After lunch, the four physicists met again in the small library.

Gell-Mann: Dear colleagues, this afternoon, we shall discuss the constants of nature. In the Standard Model of particle physics, we have at least 28 fundamental constants, but nobody knows how to calculate their values.

Newton: I only know one constant — the constant of gravity — which I introduced in 1686. What are the rest?

Haller: You know at least a second constant, the fine-structure constant, which describes the strength of the electromagnetic interaction. Also, we mentioned the constant related to quark–gluon interaction or the masses of quarks. These are also basic constants of nature. They are free parameters, which cannot be calculated: why is the fine-structure constant close to $\frac{1}{137}$ and not close to, for example, $\frac{1}{120}$? Nobody knows.

Gell-Mann: Perhaps the fine-structure constant is a cosmic accident. Right after the Big Bang, it fluctuated, and when the Universe reached a macroscopic size, it was frozen accidentally. In such a scenario, it would not be possible to calculate it.

Haller: Perhaps, that is the case.

Today, we know at least 27 different basic constants. The fine-structure constant was introduced by Arnold Sommerfeld in Munich in 1916. Unlike the gravitational constant, the fine-structure constant is a dimensionless number. We now know this number very well:

$$\alpha \approx 0.00729735257 \approx \frac{1}{137.0360}$$

The fine-structure constant describes the strength of the interaction between a photon and an electron; it is the fundamental constant of quantum electrodynamics.

Gell-Mann: Actually, the fine-structure constant is not really a constant, but a function of energy. I did the relevant calculations in 1954 together with Francis Low. The functional dependence on energy comes from quantum physics. Space is not empty, but filled with virtual electron–positron pairs. An electron in space is surrounded by these pairs. The virtual electrons are pushed away, whereas the virtual positrons are attracted. Thus, the electric charge of an electron is partially shielded. This shielding depends on the distance from the electron, that is, on the relevant energy. This process is known as vacuum polarization. The fine-structure constant, given above, is the value obtained if the relevant energy is zero.

With the LEP accelerator at CERN, the fine-structure constant has been measured at the energy given by the mass of a Z boson, that is, 91 GeV. Its experimental value is about 0.0078 or $\frac{1}{128.2}$, the value one obtains by calculation after taking into account the vacuum polarization due to all observed quarks and leptons. In other words, experiment and theory are in excellent agreement.

Haller: The strong interaction is described by the theory of quantum chromodynamics. The coupling of quarks and gluons is given by the strong-interaction coupling constant, which is also a function of energy. Due to the property called asymptotic freedom, the constant decreases as energy increases. At about 100 GeV, the strong-interaction coupling constant is about 0.12, quite small, but still 16 times larger than the fine-structure constant.

Einstein: Which fundamental constants are needed for a description of the stable matter in our Universe?

Haller: Only six. They are Newton's gravitational constant, the fine-structure constant, the strong-interaction coupling constant, the mass of an electron, and the masses of the two lightest quarks: the up quark and the down quark. The masses of quarks are important. A down quark is heavier than an up quark, and this explains why a neutron is heavier than a proton.

These six constants describe all atoms and all atomic nuclei. Other constants are needed for the description of unstable particles and the weak interaction. The weak interaction has its own interaction constant, analogous to the fine-structure constant; this interaction constant describes the interaction of weak bosons with quarks and leptons. The masses of the charged and neutral weak bosons, 80.4 GeV and 91.2 GeV, respectively, are also important for the description of the weak interaction. We take the mass of the charged weak boson as a fundamental constant; the mass of the neutral boson is, then, fixed by the weak-interaction coupling constant and the mass of the charged weak boson.

In our Universe, there are also two charged unstable leptons: the muon (mass: 105.7 MeV) and the tauon (mass: 1777 MeV). Furthermore, we have four unstable quarks: the strange quark (mass: about 0.13 GeV), the charm quark (mass: about 1.1 GeV), the bottom quark (mass: about 4.3 GeV), and the top quark (mass: about 174 GeV).

The flavor mixing of quarks is described by three angles and one phase parameter. The masses of the weak bosons are introduced by the coupling of the weak bosons with a neutral scalar particle — the Higgs particle. The mass of the Higgs particle is unknown. We do not even know whether the Higgs boson exists. I believe that there is no Higgs particle. We still do not understand how the masses of weak bosons, quarks, and leptons are generated.

In 2012, a new boson was discovered at CERN with the new LHC accelerator. The mass of this boson is about 125 GeV, but thus far, it is not known whether it is the Higgs particle or another boson — for example, an excited weak boson. Altogether, we have now 19 fundamental constants.

Einstein: You did not mention neutrinos. Since now we know that neutrinos have masses, I guess there are more constants, in particular the masses of neutrinos.

Haller: Yes, but besides the three neutrino masses, there are three mixing angles and three phase parameters. Thus, there are nine more constants, making it 28 altogether. Here is a small table of the constants:

Gravitational constant:	1
Fine-structure constant:	1
Coupling constant of QCD:	1
Coupling constant of the weak interaction:	1
Mass of the charged weak boson:	1
Mass of the Higgs boson:	1
Masses of the six quarks:	6
Mixing angles and the phase parameter of quarks:	4
Masses of the six leptons:	6
Mixing angles and the phase parameter of leptons:	6
	28

If all the quark and lepton masses are set to zero, 20 fundamental constants vanish and only six remain: the four coupling constants and the masses of the charged weak boson and the Higgs boson.

Newton: In your list, I do not find the proton mass. Why?

Gell-Mann: The proton mass is not a fundamental constant since it can be calculated in terms of the QCD coupling constant.

Newton: Strange, the proton mass can be calculated, but not the electron mass.

Gell-Mann: Yes, but we know a lot about the proton; for example, its substructure is given by the three quarks and the gluons. Nothing is known about the electron. In the Standard Model, an electron is point-like.

Newton: I think that an electron also has a substructure, but the extension of an electron is at least 1000 times smaller than the extension of a proton, because a point-like particle cannot have a mass.

Gell-Mann: Yes, probably you are right. My colleague Dick Feynman believed that the radius of an electron is about 10000 times less than the radius of a proton. Perhaps, he was right, but in this case, the substructure of an electron will be discovered with the new LHC accelerator at CERN.

Einstein: Nobody knows. We should not speculate now. But who fixes the fundamental constants? Perhaps, there exist laws of nature that have not yet been discovered, and they determine the special values of the constants. The other possibility is that the constants are accidental products of the Big Bang.

Gell-Mann: I believe that the fundamental constants are accidents. If the Big Bang were to be repeated, the constants would have other values. This would explain why nobody could calculate the fundamental constants: accidents cannot be calculated. Also, we do not know whether the constants are really constant. Perhaps, they vary very slowly in time and space.

Newton: Perhaps, my gravitational constant depends on time?

Haller: Paul Dirac proposed in 1937 that the gravitational constant is proportional to $\frac{1}{T}$ where T is the age of the Universe. This would imply that the relative change in the gravitational constant per year is about 6×10^{-11}. For a long time, such a small change could not be excluded, but now, precise experiments done by NASA using satellites have excluded such a time variation to less than 10^{-12} per year.

Let me mention again the time variation in the fine-structure constant. First, I will discuss the limits on the time variation. The best limit was derived from the remains of a natural reactor that operated about two billion years ago in Gabun, Western Africa, near the river Oklo.

In 1972, French nuclear physicists discovered that there was a natural reactor in this area. The energy output of this reactor was small, only about 100 kW, but it operated for a long time, about half a million years. By studying the remnants of this reactor, one was able to uncover details about the nuclear processes that took place about two billion years ago. Especially studied was the cross section of the reaction between the element Samarium and neutrons that produces another isotope of Samarium and a photon:

$$^{149}Sm + n \rightarrow {}^{150}Sm + \gamma$$

The cross section of this reaction is very large, and today, we know that this is due to a nuclear resonance just above the threshold. The energy of this resonance depends on the fine-structure constant. If this constant changes with time, the energy of the resonance must also change. Thus, one was able to find a limit on the time variation in the fine-structure constant. If other constants do not change with time, the limit is about 10^{-16} per year.

Fig. 9.1. *The Keck telescope in Hawaii*

I shall now mention the experiments in astrophysics. With the Keck telescope in Hawaii (see Fig. 9.1), a group of astrophysicists from Australia, England, and the United States studied the fine structure of atoms in very distant quasars (see Fig. 9.2), some of which are more than 10 billion light years away from the Earth. They discovered a very small time variation in the fine-structure constant:

$$\frac{\Delta\alpha}{\alpha} \approx -(0.54 \pm 0.12) \times 10^{-5}$$

By a linear approximation, the relative change in the fine-structure constant per year would be about 10^{-15}.

Fig. 9.2. *A model of a quasar*

Newton: But this change is larger than the limit derived from the Oklo reactor.

Gell-Mann: Remember, the limit derived from the Oklo reactor is valid only if all other constants, such as the coupling strength of the strong interaction, do not vary in time. This is probably not the case.

Einstein: Also, the linear approximation might be wrong: the fine-structure constant might have changed during the first 10 billion years after the Big Bang, but for the last three billion years, it may have been constant. Then, there would be no problem.

In the theories of Grand Unification, the coupling constants of the electromagnetic, the weak, and the strong interactions are related to one another. Thus, a time variation in the fine-structure constant would also imply a time variation in the strong coupling constant. Do we know something about this?

Haller: Well, I calculated the time variation in the QCD coupling constant to be about 40 times larger than the time variation in the fine-structure constant.

Newton: The masses of atomic nuclei depend on the QCD coupling constant. Thus, the time variation in these masses would be about 40 times larger than that in the fine-structure constant. I think this might be observed in experiments.

Haller: Yes, this is possible. Today, atomic clocks, in particular Cesium clocks, are used to measure and define time: one second is defined as the duration of a certain number — 9 192 631 770 — of the oscillations of the Cesium atom.

Theodor Haensch, a professor in Munich, compared the time measured with a Cesium clock with the time provided by a hydrogen spectrometer. The oscillations of a Cesium clock are related to a hyperfine transition, which is determined by the mass of a Cesium nucleus; in contrast, the oscillations of a hydrogen spectrometer depend only on the fine-structure constant. Thus, the time measured with a Cesium clock is different from the time measured with a hydrogen spectrometer.

I calculated that the time difference is about three Cesium oscillations per day. This means that the relative effect should be about 5×10^{-14} per year. Such a difference is large enough to be measured by Haensch. However, he did not find the effect I had predicted; instead, he obtained a new limit of about 3×10^{-15} per year. Perhaps, a time variation in the QCD coupling constant will be found in the future, but it cannot not be larger than 3×10^{-15} per year.

Finally, I would like to mention another experiment that indicates a time variation in the QCD coupling constant. Two very distant quasars that are about 12 billion light years away from the Earth have been investigated using the "Very Large Telescope" (VLT) in Chile (see Fig. 9.3). In particular, by investigating the transitions of hydrogen, it

Fig. 9.3. *The Very Large Telescope in Chile*

was observed that the ratio of the proton mass to the electron mass varies with time. If we assume that the electron mass is constant, we find that the time variation in the QCD coupling constant is 3×10^{-15} per year, which is still within the limit set by Haensch's experiment. If this experiment is correct, the masses of all atomic nuclei would change slowly with time.

It is now time for a break. Let us meet again in about 30 minutes for our last discussion, which will be about cosmology and the Big Bang.

THE BIG BANG

10

Haller: We now come to the last part of our discussion on particle physics: the field of cosmology, which is a very old science. About 5000 years ago, Chinese philosophers were already studying cosmological problems, and about 2500 years ago, Greek philosophers, in particular Thales, Plato, and Aristotle, developed the first cosmological models. But cosmology as an exact science only started in 1917, when you, Mr. Einstein, wrote an interesting paper on cosmology, after coming up with the general theory of relativity.

Einstein: Yes, I tried to construct a universe that is stable and unchanging over time. But the equations of general relativity did not allow this; the Universe must either be expanding or contracting. An expanding universe would imply that a long time ago, there was a Big Bang, the creation of the world. I did not like this since it reminded me of religion, of the creation of the world by God. But then, I was able to construct a stable universe by inventing a new term, which was later called the cosmological term, for the equations.

However, in 1922, Alexander Friedman, in Petersburg, developed a model of a dynamical universe (a universe that either expands or contracts) without a cosmological constant. In 1929, I visited Caltech, where Edwin Hubble told me that he had discovered that distant galaxies move away from the Earth at velocities proportional to their distances from the Earth. He interpreted this phenomenon as an expansion of the Universe, and I was convinced that he is right. The Universe is not stable; it is changing with time. I now regard the invention of the cosmological constant as the biggest blunder of my life.

Gell-Mann: By using modern telescopes, in particular the Hubble space telescope (see Fig. 10.1), which has been in orbit since 1990, one can observe very distant galaxies (at distances of up to about five billion light years away) to be moving away from the Earth at high speeds. With these observations, the expansion rate of the Universe today is known to a high precision, and one can estimate that the creation of the Universe,

Fig. 10.1. *The Hubble telescope*

the Big Bang, took place 13.7 billion years ago. Thus, in 1929, when you visited Caltech, cosmology turned into a real science.

Haller: George Gamow (1904–1968), who escaped from Russia in 1934 and became a professor of physics in Boulder, Colorado, studied details of the cosmic explosion. He suggested that in the beginning, the matter in the Universe was very hot, and space was filled with radiation, with photons. Today, these photons should still be around: the space between galaxies should be filled with them. Due to the expansion of the Universe, this radiation has become quite cool, and today, it should have a rather low temperature. Gamow calculated it to be about 10 Kelvin: the cosmic radiation has become a microwave radiation.

Nobody took Gamow's prediction seriously. But in 1964, Arno Penzias and Robert Wilson, two astrophysicists who worked at Bell Laboratories at Murray Hill, New Jersey, discovered the microwave radiation. They had built a special antenna that was able to receive signals from the Telstar satellites, but during weekends, Penzias and Wilson searched for signals from the sky. They found an isotropic radiation: it was the microwave radiation predicted by Gamow; the temperature of this radiation was about three Kelvin.

In November 1989, a special satellite known as Cosmic Background Explorer (COBE) (see Fig. 10.2) was put in orbit; it was able to measure the cosmic radiation very precisely (see Fig. 10.3). The temperature of the microwave radiation was measured to be 2.73 Kelvin.

The cosmic background radiation is a leftover from an early developmental stage of the Universe, and its discovery is considered an important test of the Big Bang model. When the Universe was young, before the stars and planets were formed, it was filled with a dense gas of nucleons, electrons, photons, and neutrinos. As the Universe expanded, both the gas and the radiation cooled down. Eventually, protons and electrons were able to form neutral atoms. These atoms could no longer absorb

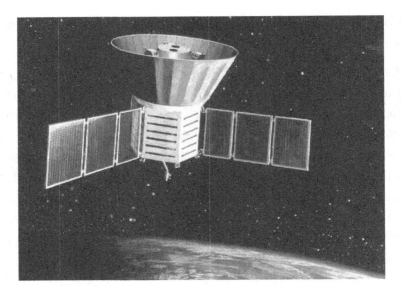

Fig. 10.2. *The COBE satellite*

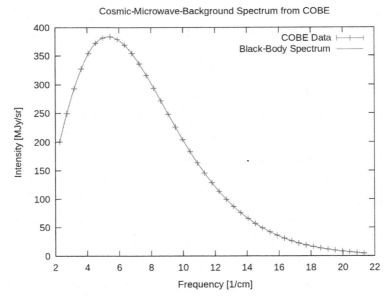

Fig. 10.3. *The result from the COBE satellite: the microwave radiation agrees perfectly with the theoretical curve given by the temperature of 2.73 Kelvin.*

thermal radiation, and the Universe became transparent, instead of being an opaque fog. Cosmologists refer to the period when neutral atoms are first formed as the "recombination epoch".

With the COBE satellite, we have also measured small fluctuations in the microwave temperature. These fluctuations can give information about the distribution of matter in the early Universe.

Newton: You mentioned that in the beginning, the Universe was also filled with neutrinos. Like photons, these neutrinos would still be around: our space would be filled with cosmic neutrinos. Is it possible to observe them?

Haller: No, this gas of neutrinos would also have a temperature of about three Kelvin. Thus, their energy is extremely low. I see no way to detect them.

Gell-Mann: A description of the development of the Universe immediately after the Big Bang requires a knowledge of the behavior of matter at very high temperatures and very high densities. Thus, we are led to the physics of elementary particles. In particle colliders, we observe collisions between electrons and positrons or between protons and antiprotons. A very high energy density is created in these collisions, just as in the Big Bang.

With our knowledge of particle physics, we can describe today the cosmic evolution shortly after the Big Bang. In the beginning, the Universe was filled with elementary particles: leptons, quarks, photons, gluons, and weak bosons. This plasma of particles then expanded and cooled down. Quarks and antiquarks annihilated each other. Then, three quarks came together to form a proton or a neutron.

Einstein: I would expect the antiquarks present to form antiprotons and antineutrons. In our Universe, there should not only exist normal matter, but also large amounts of antimatter. But this is not the case. What is the reason?

Gell-Mann: It is assumed that there was a very small asymmetry between quarks and antiquarks due to a small violation of the symmetry between matter and antimatter. In particle physics, we observe such a violation, in particular in the decays of K mesons. This symmetry violation implies that shortly after the Big Bang, there were slightly more quarks than antiquarks in the Universe.

Suppose we consider a small volume of space in which there are ten billion antiquarks. In this same volume, there are also ten billion and one quarks; the one additional quark is due to the violation of the symmetry. The asymmetry between matter and antimatter is very small, of order 10^{-10}. The ten billion quarks and the ten billion antiquarks annihilate each other, and the one additional quark remains. The matter in the Universe today consists of these additional quarks.

About one second after the Big Bang, these quarks formed protons and neutrons. Shortly afterwards, the first deuterons were created; they are bound states of protons and neutrons. About one minute after the Big Bang, the first helium nuclei were formed by the collisions of two deuterons. One can calculate how much matter is produced by these helium nuclei — about 25%. Indeed, we find today that about 25% of the matter in stars is provided by helium nuclei, which are the products of the Big Bang via a process called primordial nucleosynthesis.

About 10 000 years after the Big Bang, electrons and protons created atoms. From then on, there have been no free protons: all protons have since been bound in hydrogen atoms. These atoms are neutral, and the photons of the cosmic microwave radiation do not interact with them: the microwave radiation has become decoupled from matter. Consequently, the Universe has since become transparent. After the decoupling, the gravitational force has become relevant; due to gravity, large clusters of matter were formed. They later became galaxies or clusters of galaxies.

In 1933, the Caltech astronomer Fritz Zwicky (1898–1974) investigated the Coma cluster of galaxies (see Fig. 10.4). He measured their

Fig. 10.4. *The center of the Coma cluster of galaxies*

velocities and found they were larger than expected. He concluded that the Coma cluster could not be stable: the galaxies would fly away soon because their velocities were too large.

But this was not the case. Zwicky came to the conclusion that besides the normal matter in the Universe, there must be dark matter, which can only be observed through its gravitational effects, in the Coma cluster. The Coma cluster can only be stable if there is more dark matter in the cluster than the normal matter seen in the stars there — about 84% of the matter would have to be dark matter.

Zwicky is right. Around 1980, dark matter was also discovered inside galaxies, including our own. But up to now, we do not know what dark matter is. It must probably consist of neutral, stable particles. The only neutral particles we know that are stable are the neutrinos, but their masses are too small. Hence, neutrinos cannot be dark matter.

With the new LHC accelerator at CERN, we hope to find heavy, neutral and stable particles that could be the constituents of dark matter. But these particles might be very heavy, and in that case, they would not be produced in the experiments at the LHC. It is really frustrating that we do not know what dark matter is.

Newton: Let us forget about dark matter. It might remain a mystery for the next hundred years. I would like to understand why there was a Big Bang. Today, we have in the Universe, a lot of matter, thus, a lot of energy. Where did this energy come from?

Haller: I wish I knew, but nobody has the answer. However, your question might be related to another question. In the observed part of our Universe, cosmic radiation is very homogenous. This is a puzzle since the Big Bang is a gigantic explosion that is certainly not homogenous, and so, the microwave radiation should not be as homogenous as it is.

Einstein: In 1917, I introduced a cosmological constant in my equations of general relativity, but in 1929, I removed this constant since it was not needed for an expanding universe. Perhaps, this constant is present, and with it, one might understand these problems better.

Haller: You might be right. Your constant was recently reintroduced by particle physicists. In the Standard Model, one needs a scalar field often called the Higgs field in order to generate the masses of the weak bosons. This field has something called a vacuum expectation value; that is, this field is non-zero in a vacuum. This implies that a vacuum is not empty, but contains energy. The same is true if your cosmological constant is present. Thus, the Higgs field produces a cosmological constant, with a caveat that this constant might change in time.

Your constant might have played an important role in the Big Bang. Shortly after the Big Bang, your constant was very big, and the Universe expanded very rapidly via a process called "inflation". As a result of inflation, a very small volume about the size of an atomic nucleus expanded very rapidly to a volume as big as a tennis ball. Because of this inflation, the Universe is very homogenous and isotropic.

At the end of the inflation, a cosmic phase transition occurred: the energy provided by your constant was released, and the Universe was suddenly full of energy. Quarks, leptons, and their antiparticles were produced. This release of energy is the actual Big Bang.

After the inflation, the Universe continued to expand. One would expect the expansion rate to go down due to gravity. However, thousands of supernovas were recently investigated, and it turns out that the expansion of the Universe is, in fact, accelerating — a behavior one would expect if there is a small cosmological constant. Some physicists attribute the expansion of the Universe to "dark energy", but nobody knows what it means, unless it is just your cosmological constant.

Einstein: It seems that I may have abandoned the cosmological constant too early. I am glad to hear that there might be such a constant.

Thus far, we have interpreted the Big Bang as the cosmic explosion that created our Universe. But now, I think that our Universe is just one of many similar universes: our world might be a multiverse, containing many — perhaps, infinitely many — universes. In this case, there might be many Big Bangs in this multiverse. The world might be similar to a big fireworks, and there are cosmic explosions continuously. We do not experience this since we live in a quiet part of an old universe.

Gell-Mann: Probably, this is right. But I think the other universes have different fundamental constants. We live in a universe in which the fine-structure constant is close to $\frac{1}{137}$, but in other universes, it may be $\frac{1}{120}$, for example.

Newton: And we live in a universe in which the up quark's mass is accidentally smaller than the down quark's mass. There is no scientific reason why this is the case: it is simply an accident.

Gell-Mann: Yes, and in most universes, no life exists. Life is possible only if the fundamental constants have exactly the values we observe in our Universe. If in another universe, the fine-structure constant is $\frac{1}{120}$, for example, life is not possible.

If the up quark's mass is larger than the down quark's mass, a proton decays into a neutron, and again, there is no life. Perhaps, there are infinitely many universes in the multiverse, and in only one universe — our Universe — there exists life.

Newton: I do not like these speculations about other universes. We can never observe these universes. It is impossible for us to do experiments in another universe. Speculations about other universes should not enter the field of physics: it is some kind of metaphysics, perhaps, suitable for philosophers, but not for us; we are real physicists.

Gell-Mann: Yes, I agree with you. Let us forget the other universes. We have now arrived at the frontier of the research in cosmology. How long it will take until a major new insight is discovered is unclear, but cosmology and particle physics will remain to be interesting sciences in the foreseeable future.

Einstein: I think it will take a lot of time until new insights are found. What is "dark matter"? What is "dark energy"? Nobody knows, and progress is only possible if these questions are answered. Newton and I cannot take part in this research, but we wish you success in the near future. Perhaps, we will once again have the occasion to discuss the new results in cosmology. But now, let us end our discussion since it is quite late, and we shall have our dinner in the Athenaeum.

THE END

CHAPTER

11

The next morning, the four physicists met for breakfast in the Athenaeum. They all had the typical American breakfast: two eggs, bacon, and hash browns.

Einstein: Our discussion is now complete. Adrian, Isaac, and I will fly back to New York in three hours' time. Isaac will then fly to London, and Adrian, to Zurich.

Newton: Our discussion was very interesting, and I liked it a lot, many thanks.

Gell-Mann: I shall stay here at Caltech for a few more days. Then after that, I shall go camping for a week with my girlfriend in the Anza-Borrego Desert State Park, east of San Diego. Afterwards, I will return to Santa Fe.

Haller: I shall be back in Bern soon, but in a week or so, I will visit CERN to see whether the LHC has produced any interesting results. We should meet again in about three years' time for a discussion about the new physics brought about by the LHC.

They returned to their rooms, and an hour later, they met at the reception to wait for a taxi to the airport.

***

A stewardess woke Haller up as the plane began its descent to Los Angeles International Airport. Haller realized that he had been dreaming: he did not meet Einstein or Newton.

He could clearly see the city of Pasadena, in particular the Caltech campus. Soon, the airplane landed, and Haller took a taxi to Caltech. In the Athenaeum, he was once again able to use the Einstein room. Then, he went to the restaurant, where he met Murray Gell-Mann.

The next morning, Haller went to the Lauritsen laboratory, where he met his secretary, Carol. She showed him the office where he would be able to work during the following days. It was Richard Feynman's old office.

Short Biographies
of the Physicists

Albert Einstein

Albert Einstein was born on March 14, 1879, in Ulm, Germany, and to Hermann Einstein, a merchant, and Pauline Einstein, Hermann's wife. In 1880, the family moved to Munich, where Hermann started a company called *Elektrotechnische Fabrik Einstein & Cie*, which produced electrical appliances.

Albert was especially interested in mathematics and science. He liked to read books on science, such as Euclid's "Elements", and books on science fiction. In 1884, he started learning the violin. In 1888, he went to study at Luitpold Gymnasium[a] in Munich.

In 1894, Hermann's company went bankrupt. Consequently, the Einstein family moved to Pavia, Italy, without Albert, who remained in Munich to complete his studies at Luitpold Gymnasium. However, after a few months, he left the school and followed his family to Italy.

He applied to study at Swiss Federal Polytechnic in Zurich, but he failed its entrance examination. In order to obtain a high-school diploma, Einstein attended Aargau Cantonal School, Aarau, Switzerland. In September 1896, he received the Swiss Matura and subsequently studied at the Swiss Federal Polytechnic.

Einstein graduated from the university in 1900 with a diploma in physics and mathematics. He applied for assistant positions at several universities in Switzerland, but was unsuccessful. After working for some time as a teacher, he was hired by a patent office in Bern.

As a student in Zurich, Einstein met Mileva Maric, a Serbian student, who was also studying physics. They married in January 1903. From October 1903 until May 1905, they lived in Bern at Nr. 49 in the Kramgasse. Today, this house is a small museum.

In the apartment in the Kramgasse, Einstein wrote his famous papers, which were published in 1905. Notably, the paper on electrodynamics of moving bodies forms the foundation of special relativity.

In 1909, Einstein was appointed as an associate professor of theoretical physics at a university in Zurich. In 1911, he moved to Karl–Ferdinand University, Prague. One year later, he became a professor at Federal Polytechnic, Zurich. Subsequently, he was appointed as the director of the Kaiser Wilhelm Institute for Physics, Berlin, and as a professor at Humboldt University at the beginning of 1914.

---

[a] Today's Luitpold Gymnasium is a different school from Albert Einstein's *alma mater*, which is today's Albert Einstein Gymnasium.

In 1919, Albert Einstein divorced his wife Mileva and married his cousin Elsa Löwenthal.

In 1916, Einstein published the general theory of relativity, which describes gravitation. With this theory, Einstein calculated that light from distant stars would be bent by the gravity of the Sun. On May 29, 1919, English astronomers under the guidance of Arthur Eddington observed a solar eclipse on an island in Africa. Their observation confirmed Einstein's prediction and made him famous worldwide. In 1921, Einstein received a Nobel Prize for his explanation of the photoelectric effect.

In February 1933, while visiting the United States, Einstein decided not to return to Germany, because the country then had a National Socialist government. He became a member of the Institute for Advanced Study, Princeton, and this affiliation would last until his death in 1955. In Princeton, Einstein worked mainly on a unified theory of electromagnetism and gravity, but without success.

In 1938, nuclear fission was discovered by Otto Hahn and Lise Meitner in Berlin. Soon after that, the National Socialist government in Germany decided to build an atomic bomb. Meanwhile, Einstein wrote a letter to President Franklin D. Roosevelt to alert him of this possibility, and the United States soon started the Manhattan project to build an atomic bomb. However, Einstein was not directly involved in this project. In 1945, two atomic bombs were developed and used to destroy the Japanese cities Hiroshima and Nagasaki.

Albert Einstein died in Princeton on April 18, 1955; he was 76 years old.

Murray Gell-Mann

Murray Gell-Mann was born in New York City on September 15, 1929. His father, Arthur Gell-Mann, came from the city of Czernowitz, the capital of Bukovina, which is now a part of Ukraine. Arthur started a language school in New York.

At the age of 14, Murray Gell-Mann graduated from high school and obtained a scholarship at Yale University. He finished his studies in June 1948 with a Bachelor of Science. He then went to Massachusetts Institute of Technology, Cambridge, near Boston, for his graduate studies. His Ph.D. advisor was Prof. Victor Weisskopf. He finished his graduate studies in 1951 at the age of 21.

Afterwards, Gell-Mann went to the Institute for Advanced Study, Princeton, and started to work on particle physics. In 1952, he joined the physics department of the University of Chicago, where he collaborated with Marvin Goldberger and had close contact with Enrico Fermi. In 1955, Gell-Mann married J. Margaret Dow, who came from England and worked in Princeton University.

Gell-Mann accepted a professorship at California Institute of Technology (Caltech), Pasadena, and left Chicago in 1955. He became a full professor at Caltech at the age of 26. He had close contact with Richard Feynman. In 1993, he retired and went to the Santa Fe Institute, Santa Fe, New Mexico.

Gell-Mann's first scientific papers are related to the new elementary particles discovered in cosmic rays and with accelerators. He introduced in 1952 a new quantum number named "strangeness". With this new quantum number, he was able to describe the behavior of the new particles, which were later called "strange particles". In 1958, Feynman and Gell-Mann introduced the V-A theory of the weak interaction, which was later used in the gauge theory of the electroweak interaction. In this theory, weak currents are left-handed; parity is violated maximally.

In order to describe new hadrons, hyperons, and K mesons, Gell-Mann introduced a new symmetry based on the SU(3) group. In this symmetry, baryons are described by an octet consisting of the two nucleons, the Lambda baryon, the three Sigma baryons, and the two Xi baryons. Mesons are also described by an octet, consisting of the three pions, the four K mesons, and the eta meson. Excited baryons are described by a decimet consisting of the four Delta resonances, the three Sigma resonances, the two Xi resonances, and the Omega-minus particle. In 1969, Gell-Mann received a Nobel Prize for his work on the SU(3) symmetry.

In 1964, Gell-Mann observed that the known baryons and mesons could be interpreted as bound states of SU(3) triplet particles, which he called "quarks". A baryon is a bound state of three quarks; a meson is a bound state of a quark and an antiquark. The electric charge of a quark is non-integral: $\frac{2}{3}$ or $-\frac{1}{3}$. The SU(3) symmetry requires that there be three different quarks, which Gell-Mann called "up" (u), "down" (d), and "strange" (s). The proton has the internal structure "uud"; the Omega-minus particle has the structure "sss".

In 1971, Gell-Mann and this author, proposed that quarks carry a new quantum number, which they called "color". This quantum number is associated with the symmetry group SU(3). Observed hadrons are color singlets. In 1972, Fritzsch and Gell-Mann, by using the color group as a gauge group, constructed a gauge-field theory — the theory of "quantum chromodynamics" (QCD).

Gell-Mann's papers on particle physics provide the foundation on which the "Standard Model of particle physics" was erected. This model is the main result of the fundamental physics in the second part of the 20th century.

Isaac Newton

Isaac Newton was born on 4 January 1643 in Woolsthorpe-by-Colsterworth, near Grantham, in the county of Lincolnshire, about 150 km north of London. He went to school in Grantham. In 1661, Newton began his studies at the University of Cambridge.

Soon after Newton obtained a degree in August 1665, the university was temporarily closed as a precaution against the Great Plague, and Newton went home for two years. There, he studied in detail differential calculus and gravitation. In 1667, he returned to Cambridge.

His professor Isaac Barrow was about to retire, and he recommended Newton to be his successor. In 1669, Newton was appointed as the Lucasian Professor of Mathematics; he was rather young then, only 26 years old.

In 1687, Newton published his famous book *Philosophiae Naturalis Principa Mathematica* ("Mathematical Principles of Natural Philosophy"). In that book, Newton introduced the law of gravitation and the three laws of motion. He showed that Kepler's laws of planetary motion could be derived from his law of gravitation. He introduced the concepts of absolute time and absolute space, which were regarded as the basic principles of physics for the next 200 years until 1905, when Albert Einstein replaced them with the principles of relativity.

Newton was convinced that light consists of small particles. In 1675, he introduced the ether as the medium for light. His theory of light disagreed with the popular wave theory of light introduced by Christiaan Huygens.

In 1696, Newton gave up his professorship and moved to London. He started to work on history, alchemy, and theology. He accepted a position in the Royal Mint and in 1699, became the Master of the Royal Mint, a position he held for the last 30 years of his life. He became the President of the Royal Society in 1703.

Isaac Newton died in London on 31 March 1727 and was buried in Westminster Abbey.

GLOSSARY

Alpha particle: The atomic nucleus of helium, a bound state of two protons and two neutrons

Antiparticle: Every particle has its own antiparticle. For example, the antiparticle of the electron is the positron. The electric charge of an antiparticle is opposite that of its particle.

Asymptotic freedom: A property that causes the coupling parameter in quantum chromodynamics to decrease as energy increases
At very high energies, the coupling between quarks is very weak.

Athenaeum, The: The guesthouse at Caltech, Pasadena

Baryon: Any particle that is a bound state of three quarks
It has a half-integer spin and takes part in the strong interaction. An example of a baryon is the proton.

Beta decay: The decay of a neutron or that of an atomic nucleus by the weak interaction into a proton or another nucleus with the emission of an electron or a positron

Boson: Any particle with an integer spin
Examples of bosons are pions and photons. The name is derived from that of the Indian physicist Satyendra Nath Bose.

Cabibbo angle: A mixing angle used to describe the weak decays of the hyperons
It was introduced by Nicola Cabibbo.

Cabibbo–Kobayashi–Maskawa matrix (CKM matrix): A 3×3 unitary matrix that contains information on the strength of flavor-changing weak decays

California Institute of Technology (Caltech): A private research university in Pasadena, California

CERN: A research center for particle physics near Geneva, Switzerland
It stands for *Conseil Européen pour la Recherche Nucléaire* (which is today's *Organisation européenne pour la Recherche nucléaire*).

Color charge: A property of quarks and gluons that is used in the theory of quantum chromodynamics (QCD)
Quarks have three different colors, usually denoted as "red", "green", and "blue".

DESY: A research center for particle physics in Hamburg, Germany
It stands for *Deutsches Elektronen-Synchroton*.

Dirac equation: A relativistic wave equation introduced by the British physicist Paul Dirac in 1928
It describes the fields of elementary, spin-$\frac{1}{2}$ particles (such as electrons) using four complex numbers.

Electroweak interaction, The: The unified description of the electromagnetic and the weak interactions

Fermi National Accelerator Laboratory (Fermilab): A research center for particle physics just outside Batavia, Illinois, near Chicago

Fermion: Any particle with a half-integer spin
Examples of fermions are electrons and protons. The name is derived from that of the Italian physicist Enrico Fermi.

Fine-structure constant, The: The coupling constant of the electromagnetic interaction
It is a dimensionless number close to the inverse of the prime number 137

GeV: Giga-electronvolt, a unit of energy
One gigaelectronvolt is one billion electronvolt. The mass of a proton corresponds to about 0.94 GeV.

Glue meson: Any particle that is a bound state of two gluons and is in a color-singlet state
It is expected in QCD that such particles exist, but they should mix strongly with quark–antiquark mesons. Thus far, no glue mesons have clearly been identified in an experiment.

Gluon: The particle that is exchanged between quarks and generates the strong force in QCD

Grand Unification, The: The unification of the strong and the electroweak interactions
Thus far, the Grand Unification is a hypothesis: current experiments cannot test it since the energy at which it becomes relevant is too high.

Hadron: Any particle that is made of quarks and interacts via the strong interaction
Examples of hadrons are protons and pions.

Higgs mechanism, The: A mechanism that is used to introduce the masses of the weak bosons in the electroweak gauge theory
The name is derived from that of the Scottish physicist Peter Higgs.

Hyperon: Any baryon that consists of at least one strange quark, but no charm quarks or bottom quarks

Inflation: The fast expansion of the Universe immediately after the Big Bang

Isospin symmetry: see SU(2) symmetry

Kamiokande: A big particle detector located in a mine in Japan
It is used for the investigation of neutrinos and that of the proton decay.

Kavli Institute for Theoretical Physics (KITP): A research institute of the University of California, Santa Barbara

Large Electron–Positron Collider (LEP): To date, the most powerful accelerator of leptons ever built
It is located at CERN.

Large Hadron Collider (LHC): The world's largest and highest-energy particle accelerator
It is located at CERN.

Lepton: Any elementary, spin-$\frac{1}{2}$ particle that does not interact via the strong interaction
There exist six leptons: the electron, the muon, the tauon, and their associated neutrinos.

Meson: Any particle that is a bound state of one quark and one antiquark
It has an integer spin and takes part in the strong interaction. An example of a meson is the pion.

Muon: A charged lepton, like the electron, but with a mass that is about 200 times larger than the electron mass
It is unstable and decays into an electron and two neutrinos.

Neutral current: The current coupled to the neutral Z boson

Neutrino: Any neutral lepton
There exist three neutrinos: the electron neutrino, the muon neutrino, and the tau neutrino. Each neutrino is associated with a charged lepton.

Neutrino oscillations: The neutrinos produced via the weak interaction are not pure mass eigenstates, but mixtures of mass eigenstates. A moving electron neutrino can change into a muon neutrino and vice versa. These changes are called "neutrino oscillations".

Neutron: A neutral, unstable baryon (a bound state of three quarks, like the proton) that decays into a proton, an electron, and an antineutrino
The neutron mass is about 940 MeV.

Omega-minus particle: A hyperon of spin $\frac{3}{2}$ that is a bound state of three strange quarks

Parity: A parity transformation is the inversion of one (or more) of the three space dimensions. The strong and electromagnetic interactions are invariant under this symmetry, but the weak interaction violate it.

Photon: The particle of light and of the electromagnetic radiation
It has a spin of one.

Physical Review Letters (PRL): A peer-reviewed, scientific journal

Pion: The lightest meson (a bound state of a quark and an antiquark)
There are three pions: two charged ones and a neutral one. The pion mass is about 140 MeV. Pions are unstable: a charged pion decays via the weak interaction mostly into a muon and a neutrino; a neutral pion decays via the electromagnetic interaction into two photons.

Proton: The only stable baryon and a particle with a positive electric charge, the proton is a bound state of two up quarks and one down quark. The atomic nucleus of a hydrogen atom is a proton. The proton mass is about 938 MeV.

Proton decay: The decay of a proton — for example, into a positron and a photon

In Grand Unfication Theories, one expects the proton to be unstable, but thus far, no proton decay has been observed.

Quantum chromodynamics (QCD): The gauge theory of the strong interaction

It describes the interaction of quarks and gluons.

Quantum electrodynamics (QED): The gauge theory of the electromagnetic interaction

It describes the interaction of charged particles and photons.

Quark: The constituent of baryons and mesons

There exist six different quarks in our Universe. The lightest quarks (the up and the down quarks) are the constituents of atomic nuclei.

Quark–lepton family (or generation): A group of particles consisting of the three colored up-type quarks, the three colored down-type quarks, a charged lepton, and a neutral lepton

For example, the first family consists of the three colored up quarks, the three colored down quarks, an electron, and an electron neutrino.

Quasar: The energetic nucleus of a very distant galaxy

It surrounds a massive black hole.

Schrödinger equation: A differential equation that describes the dynamics of the wave functions in quantum theory

Spin: The intrinsic angular momentum of a particle

It can have the values 0, $\frac{1}{2}$, 1, and so on. Pions have spin 0, electrons have spin $\frac{1}{2}$, and photons have spin 1.

Stanford Linear Accelerator Center (SLAC): A research center for particle physics in Palo Alto, California
At SLAC, the interaction of high-energy electrons and of atomic nuclei were investigated.

Sudbury Neutrino Observatory (SNO): A neutrino observatory located 6800 feet (about 2 km) underground in a nickel mine in Sudbury, Ontario, Canada

SU(2) symmetry: Also known as isopin symmetry
It is slightly broken and describes the dynamics of nucleons. The nucleons (namely, the proton and the neutron) are a duplet of SU(2).

SU(3) symmetry: An extension of the isospin symmetry
It is strongly broken and describes the dynamics of baryons and mesons. The lightest baryons and mesons are octets of SU(3).

Tauon: The third-generation charged lepton with a mass of about 1.78 GeV

W boson: The charged mediator of the weak interaction
Its mass is about 81 GeV.

Yang–Mills theory: An extension of the electromagnetic interaction proposed by Wolfgang Pauli in 1953 and by Chen Ning Yang and Robert L. Mills in 1954
In the theory, the gauge group U(1) of electromagnetism is replaced by a larger group. Examples of Yang–Mills theories are the electroweak theory (gauge group SU(2) × U(1)) and the theory of quantum chromodynamics (gauge group SU(3)).

Z boson: The neutral mediator of the weak interaction
Its mass is about 91 GeV.